Introduction to
Wireless Local Loop

For a complete listing of the *Artech House Mobile Communications Library*,
turn to the back of this book.

Introduction to Wireless Local Loop

William Webb

Artech House
Boston • London

Library of Congress Cataloging-in-Publication Data
Webb, William E.
   Introduction to Wireless Local Loop / William Webb
      p.   cm. — (Artech House mobile communications library)
   Includes bibliographical references and index.
   ISBN 0-89006-702-3   (alk. paper)
   1. Wireless communication systems.   2. Telecommunications—Switching
   systems.   I.   Title.   II. Series Artech House telecommunications library.
   TK5103.2.W43   1998
   384.5'35—dc21   97-49061
                                                                                          CIP

British Library Cataloguing in Publication Data
Webb, William
   Introduction to wireless local loop. — (Artech House mobile communications library)
   1. Wireless communication systems
   I. Title
   621.3'82

   ISBN 0-89006-702-3

Cover design by Jennifer Stuart

© 1998 ARTECH HOUSE, INC.
685 Canton Street
Norwood, MA 02062

International Standard Book Number: 0-89006-702-3
Library of Congress Catalog Card Number: 97-49061

10 9 8 7 6 5 4 3 2 1

Contents

v

Preface

L IKE MANY IN THE FIELD of wireless local loop, I gained my expertise in the area of cellular and cordless radio. On moving into the wireless local loop environment, my first reaction was to seek a good reference work that would help me understand the concepts and issues behind it. No such work appeared to exist, although many colleagues acknowledged the urgent need for one. This book is an attempt to integrate the knowledge I have gained through a wide range of sources into an introductory text on wireless local loop.

The intention here is to provide a reference that those new to the area of wireless local loop can consult to understand what it is all about, and to be able to differentiate the issues that will be key to their needs. The aim also is to provide sufficient background information that the book has value as a reference when particular parameters or methods are required.

The intention has not been to produce a scholarly tome, replete with complex equations and exhaustive reference lists. Those coming to this area are unlikely to have the time or inclination to consult such a text.

Further, most of the theory is identical to that already widely published for mobile radio systems, and it seems unnecessary to repeat it here. Instead, the book deliberately has been kept to a length that allows busy executives to read it quickly and easily. Key issues are explained intuitively, rather than mathematically, allowing this book to be of use outside scholarly environments. Indeed, the focus on pragmatic and practical issues rather than on theoretical understanding makes the book appropriate for those who have to work with real deployments of wireless local loop networks.

This book was written at a time when the first wireless local loop networks were only just being deployed. The industry is still very much in its infancy, and there are many lessons still to be learned. Every effort has been made to speed the production of the book so that it is not out of date before it is published, but certainly some of the issues relating to specific technologies will require updating only a year or so from the date of publication. The intention is to issue revisions as the technology advances.

This book is divided into five parts. Part I sets out the competitive scene, including the role of wireless local loop among fixed, cable, cellular, and cordless networks. Part II details the role of wireless local loop in a range of environments and the economics relative to fixed networks. Part III provides a short technical background, including propagation and access techniques for those who are interested, while Part IV looks in detail at the different technologies available and the means of selecting among them. Finally, Part V details the experience gained to date in deploying wireless local loop systems.

Acknowledgments

A book is never the work of a single individual. Much of the experience reported here has been gathered by my colleagues at Netcom Consultants; in particular, Mark Cornish, David De Silva, Robert Lesser, Don Pearce, and Steve Woodhouse provided valuable input. I also have gained much from previous colleagues, specifically, Professor Raymond Steele, who taught me the principles of radio communications and how to write and who gave me the enthusiasm to embark on the project of writing a book. Finally, thanks to my wife, Alison, who supported the project even though it meant less time together.

Part I

Setting the Scene

1

What is Wireless Local Loop?

I N PRINCIPLE, *wireless local loop* (WLL) is a simple concept to grasp: it is the use of radio to provide a telephone connection to the home. In practice, it is more complex to explain because wireless comes in a range of guises, including mobility, because WLL is proposed for a range of environments and because the range of possible telecommunications delivery is widening.

Figure 1.1 is a simplistic diagram of the role of WLL in the world. In a simple world, a house is connected to a switch via first a local loop, then through a distribution node onto a trunked cable going back to the switch. Historically, the local loop was copper cable buried in the ground or carried on overhead pylons, and the trunked cable was composed of multiple copper pairs. WLL replaces the local loop section with a radio path rather than a copper cable. It is concerned only with the connection from the distribution point to the house; all other parts of the network are left unaffected. In a WLL system, the distribution point is connected to a radio transmitter, a radio receiver is mounted on the side of the house,

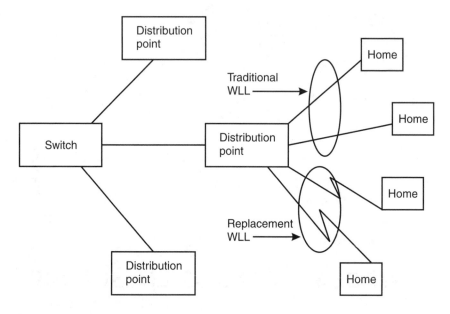

Figure 1.1 The role of WLL

in much the same manner as a satellite receiver dish, and a cable is run down the side of the house to a socket inside the house. The socket is identical to the one into which users currently plug their home telephones. Hence, apart from a small receiver on the side of their house, the home subscriber does not notice any difference.

Using radio rather than copper cable has a number of advantages. It is less expensive to install a radio than to dig up the road, it takes less time, and radio units are installed only when the subscribers want the service, unlike copper, which is installed when the houses are built. That begs the question as to why copper was ever used. As will be seen, it is only in the last five years that advances in radio technology have brought the cost of radio equipment below the cost of copper cabling. Hence, since around 1992, WLL has steadily become the most appropriate way to connect subscribers.

Most readers already will have a line to their homes, provided by copper, so they might think that WLL has arrived too late. That is not true. As will be seen, over half the world does not have a telephone line, so the market is huge. Further, in countries that do have lines, WLL is being used by new operators to provide competition to the existing telephone company. So although you may already have a copper line, you may decide to change to a WLL line.

Others may ask, Why bother with WLL, connecting the switch to the house, when cellular already connects the switch to the subscriber and provides mobility as well? That issue is discussed in more detail later, but, briefly, cellular is too expensive and provides insufficient facilities to represent a realistic replacement for a wired phone. Special systems are required, tailored to WLL deployment and that can provide a realistic alternative. Perhaps by the year 2005, cellular and WLL will have merged into one system. Until then, WLL requires specialized technology to deliver high-quality services for a price competitive to wireline.

The rest of this book is devoted to explaining the remarks in this opening section in more detail, so the reader is aware of the roles and limitations of WLL and is convinced about some of the sweeping statements made.

1.1 Why a book on wireless local loop?

Most telecommunications professionals by now have noticed the dramatic rise in interest in a topic called WLL. From its hardly noticed introduction in the early 1990s, WLL now commands numerous conferences, analysis' reports, and hype. It is also surrounded by more confusion and lack of information than most communications topics in recent years. At least the world of mobile communications has its standards, even if those standards are different in the United States and Europe; WLL has no standards to speak of. At least mobile communications knew that its main market was voice; WLL systems are suggested for voice, data, Internet access, TV, and other new applications by the day. At least mobile communications was a fairly simple proposition to put to consumers; a wireless phone that looks like an ordinary phone is bound to prompt the question "Why?" from most customers.

There is no shortage of information on WLL. Type "wireless local loop" into your Internet search engine, and you will get more than 10,000 references. Ask a conference organizer for documentation from WLL conferences, and you will need several helpers to carry it all away. Why, then, even more information on WLL?

The problem with the information currently available is that it all is essentially promotional material. As you will see later in the book, many large industrial concerns are vying for a share of a market whose size could rival that of the global cellular industry. The success of each competitor depends on its convincing the world telecommunications community that its product or service is the best. The claims and counterclaims are confusing, often misleading, and sometimes completely incorrect. It is against that background that the need for an objective assessment becomes apparent.

There are many things a book can do. It is a good opportunity to put a topic into context, to provide background information, and to analyze important issues in detail. There also are things a book cannot do. The time taken to write, edit, print, and distribute a book means that, at best, it will be at least six months out of date by the time of publication. Hence, in this book, there is no attempt to analyze comprehensively all the offerings from different manufacturers; such information would be out of date before the book was even in print. Instead, the different techniques and approaches adopted are analyzed to provide general guidelines within which each technology can be considered.

It also is worth remembering that WLL is still in its infancy. Hence, this book cannot provide authoritative and final answers based upon exhaustive experience. The information gathered from a range of sources will be proved in some cases; in other cases, it will be provisional.

1.2 Acronyms and terminology

Like most other technical fields, WLL abounds in acronyms. A full list of acronyms is provided at the end of this book, but before venturing into such a list, the issue of what WLL actually is needs to be addressed.

The term "wireless local loop" is the concatenation of the terms "wireless" and "local loop." Few readers will have any problems understanding "wireless"; it is, of course, the transfer of information without the use of wires, which implicitly means using the electromagnetic spectrum and typically the part of that spectrum known as radio. "Local loop" is the part of the telecommunications network that connects homes to the nearest local switch or distribution point. Thus, "wireless local loop" is the use of radio to make a connection from some local switching or distribution point in the fixed network to a number of houses. The reason why anyone would want to do that is introduced in Chapters 2 and 3.

The abbreviation WiLL, used by the major manufacturer Motorola, means the same thing. The UK government decided to rename the term *radio fixed access* (RFA). Other similar terms in widespread use include *fixed radio access* (FRA) and *radio in the local loop* (RLL). WLL, however, appears to be the most widely used acronym; hence, that is the term used throughout this book. (WiLL is used only to indicate Motorola's WLL radio technology.)

1.3 How to read this book

Few professionals have the time or the inclination to read an entire book. Many readers will come to this book with substantial prior knowledge and their particular agenda as to what they want to gain from the time and effort involved in reading it. With that in mind, this book has been assembled on a compartmentalized basis, allowing readers to read only the parts that are of interest to them. Further, the book has been kept relatively short. It is tempting to fill a book with all the background information that might be relevant, but here a conscious effort has been made to keep the information presented to a minimum and to provide a few choice, widely available references, to allow readers to gain the maximum from their efforts to read the book.

The book is divided into five discrete parts, as follows:

- Part I provides some background information to the telecommunications environment and introduces the concepts of convergence and access technologies. Part I places WLL in a wider and rapidly changing telecommunications environment. It should be of interest to all readers, except perhaps corporate strategists who are already well versed in its concepts.

- Part II explains why in the last 10 years the concept of using wireless in the local loop has emerged to challenge the traditional approach of laying copper cable. It first looks at the needs of the different parts of the world and then discusses, in general terms, the relative economics of wired and wireless interconnection. It finishes with some market forecasts for WLL. Part II will be of most interest to readers new to the world of WLL. Those who have visited a few WLL conferences will be familiar with the material and can pass it over.

- Part III provides some technical information on wireless. A good understanding of many of the key parameters, such as range and capacity, both of which have a critical effect on network economics, can be reached only with a little technical background. Part III looks at radio propagation and radio systems and considers a key debate: whether *code division multiple access* (CDMA) or *time division multiple access* (TDMA) forms the most appropriate access technique. Readers prepared to take on trust later claims relating to technology can skip the chapters in this section.

- Part IV is concerned with selection of the most appropriate technology. In a world where there are no standards and over 30 competing systems, such a choice is both complex and critically important. The different technologies available are introduced and impartially evaluated. Chapter 13 provides guidance on the process of making a selection. Everyone involved in technology selection should read the chapters in this section.

- Part V moves away from theory and technology and looks at the deployment of a WLL system. It describes license application, examines selection of a service offering, and develops the business case. It finishes by looking at network deployment and customer care issues. Part V will be of interest to all those involved in managing the deployment and operation of a WLL network.

Part I now continues with two more chapters: Chapter 2 introduces the converging world of telephony, TV, and computing, a convergence that is dramatically shaping WLL offerings. Chapter 3 looks at the different access technologies with which WLL will be competing for a market share.

2

The Converging World of Telephony, TV, and Computers

THE WORLD OF TELECOMMUNICATIONS used to be much simpler. For most of its 100-year history, the only option was getting a fixed line from the national (monopoly) operator. A variable was introduced in the mid-1980s, when mobile phones became a viable service but only as an addition to a fixed phone—quality was too poor and cost too high to use a mobile phone when a fixed phone was an option.

The world of broadcasting also was simple. TV signals were broadcast terrestrially in the UHF frequency band (typically 400 to 800 MHz), and anyone with a rooftop antenna and a TV receiver could receive them. Then along came cable and satellite, both offering a much higher number of channels. For our purposes, satellite systems can be considered similar to terrestrial systems: both use radio spectrum to broadcast their signals. Cable, however, is different. High-bandwidth wired connections to customers opened new opportunities.

The world of computing also was happily doing its own thing. Few computers were connected to anything other than internal networks, and all computer connections were by wire, not by radio. Corporate networks were connected with dedicated connections, and home users were not connected into anything.

Then a number of developments started to blur the boundaries between those different areas:

- Telecommunications operators started to look at how to send more information to users, opening the door to the possibility of providing TV and computing information via telecommunications channels.

- Broadcast providers started to provide telecommunications in addition to broadcasting, for example, cable operators began providing telephony services.

- Connecting a computer to the Internet via a telecommunications medium rapidly gained popularity.

Today, almost every delivery medium aims at providing telephony, broadcast entertainment, and computer services. That is the convergence that many in the field have talked about for so long. To labor a point and for those for whom "convergence" is nothing more than a useful term to use in conversations with clients, consider this. Your local cable operator will offer you broadcast TV, telephony, and Internet access all through the same line. By clicking an icon on an Internet Web page, you can automatically call the company whose Web page you are viewing. Based on your questions, the company can change the Web page you are viewing or help you download a video clip, which you can then view on your *personal computer* (PC) or as a short program on your TV set. Such service is available today and is gaining rapidly in capability and ubiquity.

The effect is that the traditional boundaries are lost. Phone calls now link computers. Broadcast TV now provides Internet data. Internet links provide voice conversation. A WLL operator in all but the least developed countries that ignores the evolution of convergence is unlikely to survive. This chapter examines those trends and summarizes the new competing forces.

2.1 Telecommunications

In the major countries in the world, most households are linked via a copper line into the *public switched telephone network* (PSTN), the world's largest manmade network, which allows anyone to phone anywhere in a few seconds—anyone, that is, connected to the PSTN and anywhere that is also connected to the PSTN. For those for whom wired telephones are an integral part of everyday life, it is easy to forget that much of the world does not have telephone access. According to statistics from the *International Telecommunications Union* (ITU), more than half the world's population have never made a telephone call. Telephone penetration varies from 60% of the population in developed countries to 2% or less in undeveloped countries. The penetration figure is the total number of installed phone lines divided by the total population. In developed countries, some phones are shared among households so although the penetration might be only 60%, access to telephones is close to 100%. The dramatic difference in access to telephones is one of the key drivers for WLL and a topic that is discussed in detail in Chapter 4.

The copper wire used typically is referred to as twisted-pair copper, two thin copper wires surrounded by insulation and twisted together. Such wiring has the advantage of being inexpensive but the disadvantage that it tends to act as an aerial, radiating high-frequency signals, potentially into other twisted pairs lying nearby and causing the phenomenon known as cross-talk. However, because voice does not contain much in the way of high-frequency components, limiting the signal that the line carries to around 3 kHz prevents cross-talk. (Chapter 3 provides more technical information about twisted-pair copper; the point here is to emphasize the difficulty in sending high-frequency signals through twisted-pair copper.)

Despite its limitations, twisted-pair copper has one important characteristic: it links almost every house in the developed world to the PSTN. Not only that, it is the only link for most houses. As will be shown later, putting in more copper is extremely expensive. Those factors combine to ensure that researchers will concentrate on the means to get the most possible out of the existing twisted-pair copper.

The owners of twisted-pair connections also have a great interest in expanding what those connections can be used for. The owners are the large state telecommunications companies (e.g., BT, France Telecom, AT&T, and the so-called baby Bells in the United States) and normally are referred to as *post and telecommunications organizations* (PTOs). In the post-Thatcher and post-Reagan world of market economics, the PTOs are a prime target for government action. The PTOs typically have a monopoly of access to the customer, but governments prefer to see a competitive marketplace, one in which abuse of monopolistic power is less likely to occur. To encourage such a marketplace, governments typically both prevent the PTOs from entering new marketplaces and encourage other operators, often on preferential terms, to compete with the PTOs. The latter strategy—yet another key driver for WLL—is examined in Chapter 4. The former tactic is a key driver for convergence and is discussed below.

PTOs, while eager to grow, are restricted by government legislation from entering areas such as broadcasting and computing. They are not, however, restricted from using their existing cabling to enhance consumer choice. Hence, PTOs are keen to find ways to send video down twisted-pair copper to provide services such as *video on demand* (VOD). VOD allows a user to view a film at home without having to go to the video store to rent a tape.

PTOs also want to use their twisted pairs to provide connections to the Internet, maximizing the time that the cabling is used. If that provides the incentive for users to request a second twisted-pair connection, so much the better for the PTO.

So, from being a means whereby people can talk with each other, the twisted-pair connection rapidly is becoming a channel where a range of different media types is transmitted, including voice, video, and computer data, that is, the much vaunted multimedia world. With all the different types of information passing through the same channel, it is relatively simple to link them so that a film can be ordered through the Internet and the Internet and voice calls can work together; thus, the emergence of the PTOs and twisted-pair copper as major factors in the convergence of audio, video, and computing.

Telecommunications has a more recent addition to its transmission means: mobile communications. Cellular communications has been one of the major growth industries of the last decade; already, only 10 years after launch, cellular revenues are some 15% of total telephony revenues in most developed countries. That percentage can be expected to increase rapidly, perhaps to 50% and more over the next decade. Mobile communications cannot be ignored as a key force in telephony. The advent of mobile communications and the development of the technology and manufacturing required to deliver mobiles at low cost are key enablers that make WLL possible and economic. But more about that in later chapters.

In terms of convergence, mobile telephony is not a key driver. Only about 2% of mobile calls are data, and most of those are fax messages. The scarcity of mobile spectrum means that high-bandwidth applications such as video and computing will remain expensive for some time, as well as slow and of poor quality. Perhaps by 2005, video might become more normal on a mobile, but that will be a long time after convergence has taken place. For that reason, mobile is mentioned here only for completeness.

2.2 Broadcasting

Since the widespread advent of TV broadcasting in the 1940s, the delivery of TV signals by terrestrial transmitters that provide around four channels has been prevalent. This is known as terrestrial TV. More recently, two alternatives have appeared. Satellite TV offers 30 or more channels to anyone who installs a satellite dish. Cable TV offers 50 or more channels to anyone able to access the cable. Cable has one big advantage over the other two delivery mechanisms in that each subscriber has a dedicated link into the cable network and is thus able to transmit into the network as well as receive from it. It did not take cable operators long to realize that they could increase their revenues by offering viewers a telephone service as well as the broadcast TV service. Once a telephony service was provided, Internet access became possible, and cable operators are now looking at ways to provide higher speed Internet access.

Unlike twisted-pair copper, cable operators typically use coaxial cable (coax) to connect subscribers to the network. Coax consists of a copper central conductor surrounded by insulation and then an earth shield. The shield dramatically reduces radiation compared to twisted pair and allows cable operators to provide much higher bandwidths, typically 750 MHz, compared with the 3 kHz of twisted pair. With such bandwidth capabilities, cable can offer Internet access speeds of 40 Mbps compared to the typical 33 Kbps or so available on the twisted-pair access network.

Like PTOs, cable operators are considering converging voice, TV, and computing to allow Internet voice, Internet selection of films, and voice guidance through Internet pages.

It is not possible to venture into the world of broadcasting without hearing the world *digital*. Digital broadcasting is the hot topic of the 1990s and will affect most TV viewers by around the year 2005. Digital broadcasting works by converting the picture to a stream of binary digits and then exploiting the fact that in most cases the current frame transmitted is nearly identical to the previous frame. Hence, instead of each frame being sent, only the difference between the previous and the current frame needs to be transmitted. Coupled with plenty of other clever intelligent coding techniques, this results in a digital TV picture being transmitted in substantially less bandwidth than current analog pictures. The difference varies between a factor of around 4 to 40, depending on a range of factors outside the scope of this book; suffice it to say that digital will enable many more channels to be broadcast than analog. Digital broadcasting will appear on satellite, cable, and terrestrial transmitters during 1997 and 1998.

A digital broadcast channel has a relatively fixed capacity. However, the resources required for video transmission vary dynamically, from virtually none (e.g., during a news broadcast, when only the newsreader's lips move) to high levels (e.g., during a football game, when almost everything moves). The channel is sized for the latter situation, so transmission of the newscast has significant spare capacity. There is much talk about using that spare capacity for data download of non-time-critical information, for example, newspapers and local information. Such information will, of course, need to be processed at the TV set. Many industry observers think that TV then will offer the ability to display, edit, and request more information; in short, it will have many of the characteristics of an Internet-connected PC. Indeed, Microsoft recently announced a significant investment in providing operating systems for such a device. A TV set would almost certainly also contain a socket to plug in a telephone, allowing voice, video, and computer access through the set, in direct competition to the telecommunications provider.

Cable operators are best placed to take advantage of this trend because they already possess the return channel whereby information from the subscriber can be passed back into the network. Terrestrial and satellite broadcast may have to rely on twisted-pair connections to return information to the network, further increasing the complexity of the convergence that is taking place.

2.3 Computing

It hardly seems necessary to discuss the Internet, a topic so prevalent in everything from specialized journals to national newspapers and magazines that it is unlikely that any readers will not have a good working knowledge of this phenomenon of the 1990s. Instead, this section focuses on what the Internet means for convergence.

Suddenly, computers are no longer stand-alone devices. Many are now networked and able to draw on massive resources of information. At the simplest level, you can communicate through e-mail rather than using the telephone, an example of voice-computing integration. At another level, you can retrieve information, read advertisements, and get the news. Finally, well-specified machines can receive video clips for replay on the computer. But of course, all those activities are possible only if you are connected via a twisted pair or coax cable into the PSTN. Here we have a highly integrated scenario in which the Internet is replacing the traditional traffic over the local loop with a wider range of traffic in a different form.

Any Internet user is familiar with the slow delivery time of the Internet; files can take hours to download, and connections fail to get made. That is because the existing telecommunications networks, over which the Internet is delivered, were not designed to cope with the volumes of traffic that the Internet is increasingly generating. The convergence of computing, in which disk sizes are measured in gigabits, and telecommunications in which even the best modems manage only kilobits per second has caused some major problems and will prove key drivers for the future. New telecommunications systems need much more bandwidth but are restricted by the limits of twisted-pair connections. Wireless access may be one means to ameliorate the situation.

2.4 The new competing environment

This section looks at the implications of convergence for a prospective WLL provider. In a fully convergent and well-developed country, all the entities shown in Table 2.1 might compete to provide a more or less encompassing service to users. Each entity is shown with details of the delivery mechanism they will use and whether they will provide telephony, broadcast, or computing services.

Some of the terms in Table 2.1 have yet to be introduced. *Microwave video distribution system* (MVDS), *Integrated Service Digital Network* (ISDN), and generic *digital subscriber lines* (xDSL) are discussed in Chapter 3. For the purposes of this chapter, suffice it here to consider them as technologies capable of delivering the services listed.

Reading down the columns, it can be seen that telephony could now be provided by PTOs, WLL operators, cellular/cordless operators, and cable operators. Because of the additional services they offer, the economics of the different operators are quite different. For example, a PTO operator makes most of its money via telephony, whereas a cable operator bases its network on TV subscriptions and can provide telephony at almost no additional cost.[1]

Table 2.1
Competing Providers in a Convergent World

Entity	Technology	Telephony	Broadcast	Computer
PTO	Twisted pair, ISDN, xDSL	One and two lines	VOD	High-speed asymmetrical access
WLL operator	Wireless	Two lines	No	64-Kbps access

1. Assuming, that is, that the cable network was engineered to provide voice telephony. Some earlier cable networks require substantial reengineering to allow voice traffic to be carried.

Table 2.1 (continued)

Entity	Technology	Telephony	Broadcast	Computer
Cellular operator	Cellular and cordless	One line	No	Limited but mobility
Cable operator	Coax	One and two lines	50+ channels	High-speed symmetrical
Terrestrial broadcast	Analog and digital TV	No	5–10 channels	Some download potential
Satellite broadcast	Analog and digital	No	50+ channels	No
MVDS broadcast	Digital TV	Yes	50+ channels	High-speed asymmetrical access

An operator that can offer all types of service through one access medium should be well placed to maximize economies of scale and hence succeed in the marketplace. Cable operators come closest to that position, with the PTO next. Both operators, however, are hampered by the high cost of laying and upgrading cable; hence, their market dominance will not be as great as might have been imagined.

According to Table 2.1, the WLL operator does not look well placed to take advantage of a convergent world, with only telephony and relatively low-speed computing access capabilities. However, Table 2.1 does not provide the whole story. The WLL operator's key competitors in the convergent world will be the PTO and the cable operator. The WLL operator may even team with the terrestrial and satellite broadcasters to provide them with a return channel and increase their offering. Compared to the PTO and the cable operator, the WLL operator, as we will see in Chapter 5, is able to provide a connection for significantly less cost. Although the PTO potentially is able to offer high-speed computer access, that technology may be expensive and difficult to deploy to all areas. The same is true for cable operators, which have particular problems with the return path due to their original network design, which will be expensive to overcome. A WLL operator providing relatively good voice and Internet access on a relatively low-cost base might provide a well-targeted service for many customers.

Even better, in any particular country, not all these types of operators will be present, and there may not be a demand for all those services. Regardless, WLL operators must remember that they are operating in a world where convergence is a key driver, and failure to provide Internet access, voice, and (potentially) video is likely to undermine significantly their business case. WLL operators also are operating in a world where competitors are not just the PTOs but also the terrestrial and satellite broadcasters and the cellular operators, against which appropriate strategies must be developed.

Chapter 3 looks in more detail at the different technologies that will be used by each of the competing operators to provide access to their customers.

3

Access Technologies

WLL IS ALL ABOUT providing access from the home into the switched network. As discussed in Chapter 2, WLL is only one of a number of competing technologies that can be used to provide access. In this chapter, all the existing and proposed technologies that are, or might be, used to provide local loop access are introduced, along with a short description of their key strengths, shortcomings, and likely costs. Most access technologies merit a book in their own right; indeed, books are available on many of the topics. This chapter is intended only to provide sufficient information that WLL operators will be able to better understand the competition they face.

3.1 Access via twisted pair

3.1.1 Voiceband modems

The twisted pair can be used directly to provide voice communications. To provide data communications, it is necessary to make use of a device that converts the data signal into a format suitable for the telephone channel. Such a device is known as a modem, a shortened form of the term modulator-demodulator. A modulator takes the digital waveform and maps it onto an analog signal that looks to the telephone system somewhat like a voice signal. The demodulator reconverts the signal into an analog signal. A detailed description of telephone modems can be found in [1].

The telephone channel has a bandwidth of about 3 kHz. It also has a relatively good *signal-to-noise ratio* (SNR) of some 30 to 40 dB. That means that although only some 3,000 symbols per second can be transmitted, each symbol can contain a relatively large amount of information. Instead of representing just two different levels, as is normal in digital modulation, it could represent, say, 16 or 32 different levels. The modulation used to achieve this is termed *quadrature amplitude modulation* (QAM).

Voiceband modem standards are developed by the ITU. Standards are important in this area because the modulator and the demodulator are installed in different premises, often in different countries, and they need to know how to work with each other. The standards are updated as technical progress allows. Each is known by a number, such as V.33. The letter V is common to all modems, while the number tends to increase as each new modem is introduced. However, there are other entities that the ITU standardizes within the V series, such as interconnection arrangements. Therefore, not all V.xx numbers represent modems, and the modem numbers do not necessarily rise consecutively. An example of how the standards have progressed is shown in Table 3.1.

The most recent standards allow data rates of up to 33.6 Kbps, with the latest modem to be announced capable of rates up to 56 Kbps, depending on the quality of the channel. This recent progression reaches the theoretical maximum rate of information transfer on the band-limited twisted wire; hence, no further improvement in speed can be expected. (Subsequent sections discuss techniques that achieve much higher data

Table 3.1
Summary of Voiceband Modems

Data Rate (Kbps)	Symbol Rate (Baud)	Modulation Type	Standard
2,400	1,200	4-DPSK	V.26, 1968
4,800	1,600	8-DPSK	V.27, 1972
9,600	2,400	16-QAM	V.29, 1976
9,600	2,400	32-QAM	V.32, 1984
14,400	2,400	128-QAM	V.33, 1988
33,600	4,800	256-QAM	V.34, 1996

transfer rates, but such techniques work only when the 3-kHz band-limiting filters are removed by the PTO.)

The key advantages of voiceband modems are the following:

- The economies of scale achieved have resulted in a cost per modem of around $200 each.

- They can be connected directly to a telephone line with no need for the PTO to modify the line in any manner.

The key disadvantages are the following:

- They need a dedicated line for the time they are in use; hence, voice calls cannot be made or received on the telephone line.

- The maximum capacity is around 56 Kbps, which is relatively slow for computer data transfer.

3.1.2 ISDN

Integrated Service Digital Network (ISDN) basically is a framing format that allows data to be carried at a range of data rates across a bearer. ISDN makes use of the fact that twisted-pair cables can carry more information if the problems of cross-talk can be overcome. To provide ISDN access, the PTO first must remove filters on the line that prevent signals of bandwidth greater than 3 kHz being transmitted. There is an installation cost involved, which the user must pay. An ISDN modem is then installed at both ends of the line.

Not all lines are suitable for ISDN. Older lines, or lines over 3 km, typically cannot carry ISDN because the cross-talk with other lines is too severe or the signal attenuation too great. A test on the line is required before ISDN service can be provisioned.

ISDN is an international standard that provides a range of data rates. The lowest rate ISDN channel is 64 Kbps, with a typical ISDN deployment providing a so-called 2B + D arrangement (known as basic rate ISDN access, or BRA). There are two basic (B) 64-Kbps channels and one data (D) channel of 16 Kbps. The data channel can be used to provide signaling information, while both basic channels are in use. Hence, a 2B + D channel provides 144 Kbps. Primary rate ISDN offers 30B + 2D channels, a total of nearly 2 Mbps, but cannot be provided over twisted-pair copper; instead, new coax cable is required. Basic rate modems cost around $300 each, although prices are expected to fall significantly in the coming years. More information on ISDN can be found in [2] and [3].

The advantages of ISDN include the following:

- It is a long-established standard and a proven technology.

- It is relatively cheap and widespread in some countries.

The disadvantages include the following:

- Only a small increase in the rate is offered by voiceband modems.

- ISDN may be rapidly outdated by xDSL technology.

3.1.3 xDSL technologies

The area of digital subscriber line technologies is a relatively new one (the abbreviation xDSL refers to all the approaches to digital subscriber lines). The concept, like ISDN, is to use existing twisted pair, less any filters that may be in place, to provide significantly greater data rates through complex intelligent modems capable of adapting to the channel and removing any cross-talk that might be experienced. The term xDSL has come about to encompass a host of proposed different approaches, such as *asymmetric digital subscriber line* (ADSL), *high-rate digital subscriber line* (HDSL), *very high rate digital subscriber line* (VDSL), and doubtless more to come.

Research has shown that these technologies can offer up to 8 Mbps, perhaps more, depending on the quality of the existing twisted pair. Readers at this point may be asking themselves why on the one hand the twisted pair can provide only 56 Kbps and on the other hand the same twisted pair can achieve 8 Mbps. The reason has to do with the manner in which cross-talk is treated. Voiceband modems overcome the problem of cross-talk by ensuring that none is generated. The xDSL technologies generate significant cross-talk but employ advanced technology to cancel its effects. It is that difference in approach, enabled by advances in digital signal processing, that has allowed xDSL to make such dramatic improvements in the data rates that can be achieved.

The first of the xDSLs to appear was HDSL, which provides up to around 768 Kbps on a single twisted pair. It also can make use of a number of twisted pairs to deliver higher rate services by, for example, sending every even bit down one cable and every odd bit down another. Using up to a maximum of three twisted pairs, a maximum data rate of around 2 Mbps in both directions can be achieved with only modestly complex equipment. A major difficulty associated with HDSL is the removal of echoes from the signal, which can cause intersymbol interference. The echoes are removed by equalizers. Equalizer design is a complex topic that attempts to balance complexity and delay against performance. In HDSL, a combination of preequalization at the transmitter and equalization at the receiver is used. The preequalization attempts to transmit a signal that when received has no echoes, while postequalization removes any residual error effects.

HDSL is intended for business applications. HDSL signals can propagate only a few kilometers along twisted pairs. Most businesses, however, are relatively close to their nearest exchange, so that is not a significant limitation. HDSL typically is less suitable for residential applications, because homes may be at much greater distances from the local exchange.

After HDSL came ADSL, which provides more data in the downstream direction than in the return path. This asymmetry meets the requirements of Internet access well, where more information is passed to the home than is sent into the network from the home. By restricting the return path to lower rates, less *near-end cross-talk* (NEXT) is generated. NEXT is interference from the return signal that contaminates the received signal. Because the return signal is at a lower rate, the effect of NEXT is reduced and higher downstream rates achieved. ADSL promises to provide up to 8 Mbps downstream but only tens of kilobits per second upstream. Current trails are achieving around 1.5 Mbps downstream and 9.6 Kbps on the return path.

ADSL works by dividing the transmitted data into a number of streams and transmitting the streams separately at different frequencies. This approach is known as *discrete multitone* (DMT) in the fixed-line community; however, the technique has been used for many years in mobile radio normally known as *frequency division multiplexing* (FDM) or *orthogonal frequency division multiplexing* (OFDM). Indeed, this is the technique proposed for digital audio broadcasting and digital terrestrial TV broadcasting. For a detailed discussion of this approach, see [1]. This approach has the advantage that each transmitted data stream is narrowband and does not require equalization. The capacity of each stream can be adjusted according to the frequency response of the channel at that particular point. It also tends to improve error performance against impulsive interference, because an impulse now damages a fraction of one bit on all the channels instead of a number of sequential bits on a higher rate channel. However, additional complexity results from the need to have an echo canceler for each channel and to modulate the multiple channels onto the single telephone line.

ADSL is more appropriate for residential applications. By reducing NEXT, the range achieved is greater than that for HDSL, allowing long residential lines to carry ADSL successfully. Also, the asymmetrical signal typically is suitable for residential applications such as VOD, in which more signal is sent to the home than received from it. It is estimated that up to 70% of all residential lines in the United States could be suitable for ADSL operation.

Finally, VDSL has been proposed where *fiber to the curb* (FTTC) has been deployed. In that case, the copper run to the subscriber's premises is very short, typically less than 500m; hence, higher data rates can be supported. Using the most advanced technology proposed yet, it is suggested that VDSL could achieve data rates of up to around 50 Mbps, although that is still far from being proved. Current plans suggest 10 Mbps downstream and 64 Kbps on the return path. VDSL cannot be used in networks in which FTTC has not been implemented.

xDSL will be expensive to implement, even though the local loop will stay relatively unchanged. The PTO will need to install new optical cable from the switch to a new cabinet in the street, as shown in Figure 3.1. Modems for xDSL are predicted to cost around $500 each, although the price in the coming years will depend heavily on the success of the technology and the economies of scale achieved.

A problem with all the xDSL technologies is that the data rate that can be achieved depends on the length and the age of the twisted pair. As the length gets longer, the data rate falls. As yet, it is not clear what percentage of lines will be of sufficient quality to accept xDSL signals. Figures quoted in the industry vary from around 60% to 90%. Due to the technology's relative newness, texts on xDSL are hard to find and tend to be limited to chapters in books such as [4]. Readers who want to know more about this topic should refer to academic journals such as [5] and [6].

A summary of the twisted-pair technologies is provided in Table 3.2.

The key advantage of xDSL is the potential extremely high data rate on existing ubiquitous lines.

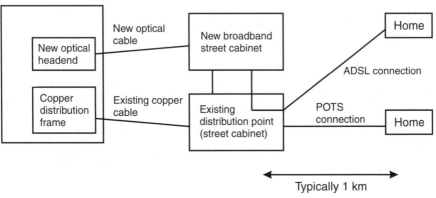

Figure 3.1 Modifications required to install xDSL technologies.

Table 3.2
Summary of Twisted-Pair Technologies

Technology	Speed Rate	Mode	Applications
Voice modems	56 Kbps	Duplex	Data comms
ISDN	144 Kbps	Duplex	Voice and data
HDSL	1.5–2 Mbps	Duplex	WAN, LAN
ADSL	1.5–9 Mbps	To user	Internet, VOD, LAN multimedia
	16–640 Kbps	To network	
VDSL	13–52 Mbps	To user	As ADSL plus HDTV
	1.5–2.3 Mbps	To network	

The key disadvantages of xDSL are as follows:

- The modems are relatively expensive.

- The technology is unproven.

- It is unlikely to work for all homes.

3.2 Access via coax

Cable operators have implemented what often is known as a tree-and-branch architecture. Figure 3.2 is a schematic representation of such an architecture.

Cable networks vary in their composition. Some networks are entirely coax, others use fiber optic in the backbone (the trunk of the tree) but coax in the branches. The latter networks are FTTC or *hybrid fiber coax* (HFC). Some postulated networks are composed totally of fiber, termed *fiber to the home* (FTTH). At present, the economics of FTTH are not favorable.[1]

---

1. FTTH has been something of a Holy Grail for PTOs and the cable industry because of the assumption that it represents the ultimate possible delivery mechanism, capable of delivering gigabits to the home. However, it has been pointed out that FTTH is akin to

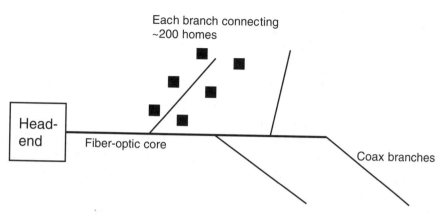

Figure 3.2 Tree-and-branch architecture.

While fiber has a virtually unlimited bandwidth (on the order of gigabits per second), coax has a bandwidth of up to around 750 Mbps in the existing installations.[2] With an analog TV picture requiring some 8 Mbps of bandwidth, that still allows numerous TV channels. With just one analog channel, some 50 Mbps of data can be transmitted using similar QAM techniques to voiceband modems. Cable, then, offers much higher capacity than even the xDSL techniques over twisted pairs.

This is not quite the whole story. For each home, there is one (or two) twisted copper pairs running from the switch right to the home. In a cable network, all homes share the backbone resource and the resource of the branch of the tree to which they are connected. To put it another

---

powering a lawn mower with a jet engine: the job might get done a little faster, but the mechanism is overkill in a situation in which coax provides more bandwidth than most users can possibly require. FTTH is both expensive and problematic in that, unlike current telephony, power cannot be supplied along with the signal. Now that the PTOs are facing competitive situations and investments increasingly have to be justified, it seems unlikely that FTTH will be implemented in the next 20 years or more.

2. Coax cable can have much higher bandwidths, up to hundreds of gigahertz, but as the bandwidth required gets higher, the cable tends to be expensive and bulky and the distance between amplifiers drops. Most cable operators have selected 750 MHz as a compromise between bandwidth and cost.

way, all homes on one branch are connected to the same cable, whereas they all are connected to their own individual twisted pair. That is fine while cable is delivering broadcast services, to be watched by many viewers simultaneously, allowing 50 or more TV channels. However, if each user on a branch wants a VOD service, then only 50 users could be accommodated on one branch, unlike twisted pair using xDSL, where as many users as needed could be accommodated. Indeed, in a typical cable network, the bandwidth per home available is only around 31 kHz, although it is unlikely that all homes would be using a dedicated downlink resource at the same time.

Such a sharing of resources causes even more problems in the return direction. Not only is the return path shared among all the users who require it, significantly reducing the capacity, but further, each user introduces noise onto the return path. The switch sees noise from across the entire network, significantly reducing the SNR and hence information content that can be received. The noise is particularly severe at low frequencies, where it often is known as ingress.

Despite all those problems, cable modems are being put on trial with a downstream capability of 30 Mbps and an upstream capability of 10 Mbps for an expected price of around $500. In summary, the advantages of cable are as follows:

- It has relatively high speed capabilities while requiring little modification to the network.

- Revenues from telephony, broadcasting, and Internet access allow the network costs to be divided across more users.

The disadvantages include the following:

- Cable penetration varies from near zero in some countries to an average of around 30% (there is near-full penetration in a few countries).

- The tree-and-branch structure may mean that twisted-pair systems start to surpass cable networks in five years or so with higher bandwidth capabilities.

3.3 Access via TV broadcast

TV signals currently are broadcast via terrestrial transmission, satellite, and cable. Cable access was discussed in Section 3.2. This section looks briefly at terrestrial and satellite broadcasting.

Terrestrial broadcasting uses about 400 MHz of radio spectrum in the UHF frequency band (typically 400 to 800 MHz in most countries). However, the same spectrum cannot be used in adjacent cells, because interference would occur. The spectrum typically needs to be divided into clusters of 11 cells to avoid interference, that is, each cell has around 36 MHz of spectrum. Some cells cover millions of users, with the result that each subscriber has only a few tens of Hertz. Thus, providing individual services for each subscriber, such as VOD, clearly is not going to be possible.

Satellite is in a similar situation. It has more spectrum, around 1 GHz in total, and does not need to divide that spectrum among different cells in the same manner as terrestrial. However, compared to terrestrial, its cells are even larger, covering over 100 million users, so again individual services for subscribers are not possible.

Neither delivery mechanism can offer a return path. Service providers for both mechanisms postulate that they might use the twisted copper of the PTOs for the return paths, but they would get limited revenue and would have difficulty truly integrating their services. So, as an access method, both leave a lot to be desired. Their main role, with the advent of digital TV, will be the delivery of broadcast data in relatively large volumes. That could speed Internet access to some of the most popular pages and provide information such as newspapers online. The key advantage of TV broadcasting is that large amounts of data can be downloaded to large volumes of subscribers.

The key disadvantages are as follows:

- It lacks a return path.

- It cannot deliver signals on a per-subscriber basis.

3.4 Access via mobile radio

3.4.1 Cellular systems

Cellular systems are available in most countries in the world. Compared to the other delivery mechanisms discussed so far, the key difference for cellular is that it provides access to a mobile terminal as opposed to a fixed terminal. However, all cellular systems suffer from capacity problems, even when they are providing voice-only traffic. Cellular systems also suffer with voice quality. Even digital cellular systems provide a voice quality that is inferior to existing fixed-line quality. The fall in quality is tolerated only because of the advantage of mobility. Few users in developed countries would consider replacing their home phones with cellular phones, so as an access method to the home, cellular is a long way from being competitive. However, in developing countries where there are no alternatives, voice quality is less important and cellular can be a viable alternative.

Some cellular operators have attempted to attack the fixed-access market using initiatives such as cut-price evening calls and high levels of indoor coverage. Invariably they have discovered that they cannot match the PTO on price and that the cost of providing excellent indoor coverage is extremely high. When recently interviewed, all the U.K. cellular operators stated that the percentage of traffic they were taking from the fixed operators was minimal, certainly accounting for less than 5% of their overall call base. Although that is likely to rise in the future, it now appears clear that cellular will not be a substitute for the fixed network but a complement when mobile.

The most successful cellular system, the *global system for mobile communications* (GSM), offers voice or data, with maximum data rates of up to 9.6 Kbps. Future enhancements to GSM might raise the maximum data rates as high as 64 Kbps and improve the voice quality, but that rate is slow compared to the other access technologies and is expensive in call charges.

Cellular systems have many disadvantages as an access technology, particularly low capacity and high cost. In developed countries, it can be expected that for some time they will continue to fulfill the role of providing mobility but will not present significant competition to the other access methods. That is not as true in developing countries, where cellular operators can use the same system to provide mobile systems and WLL systems and thus realize economies in equipment supply, operation, and maintenance.

The roles that cellular-based systems have to play in WLL networks is discussed in more detail in subsequent chapters. Fixed mobile integration also may affect the selection of an access method and is considered in more detail in Chapter 19.

3.4.2 Cordless systems

Cordless systems are similar to cellular but typically are designed for office and local area use. The differences between cellular and cordless technologies are explained in more detail in Chapters 10 and 11. The key difference in their application as an access technology is that cordless technologies such as the *digital European cordless telephone* (DECT) offer a higher data rate than cellular, up to some 500 Kbps, with lesser limitations on spectrum and hence expense associated with the call. However, because cordless technologies typically provide coverage only in buildings and high-density areas, they will be unlikely to have coverage for most access requirements. It is that lack of coverage that makes them inappropriate as an access technology. However, as we will see, when deployed as a WLL technology, cordless becomes viable as an access technology.

3.5 Access via WLL

3.5.1 Telephony-based systems

Because the remainder of this book discusses WLL as an access technology, it is inappropriate to provide too much detail at this point. Instead, the key statistics, advantages, and disadvantages are provided to allow a comparison to be made at the end of the chapter. A wide range of WLL systems provides access rates of 9.6 Kbps through to around 384 Kbps. WLL systems have the following advantages:

- There is an economic provision of service in a wide range of environments.

■ Moderate bit rates allow simultaneous voice and data in some cases.

They have the following disadvantages:

■ Bit rates will not allow high-speed connections.

■ A lack of existing infrastructure means new networks need to be built from scratch.

3.5.2 Video-based systems

As will be discussed in Chapter 12, an enhanced form of WLL technology that is on trial uses local transmitters to deliver TV signals and broadband data. Those trial systems are called *microwave video distribution systems* (MVDS) and transmit in the 40-GHz band, where around 2 GHz of spectrum has been allocated. They are highly asymmetric, with data rates around 500 MHz downstream but only around 20 Kbps return path, where it has been installed. MVDS systems have the following advantages:

■ They have very high downstream data rates.

■ They are capable of providing video, telephony, and computer data, thus maximizing revenues.

■ The cost is relatively low, in comparison with cabled alternatives.

They have the following disadvantages:

■ Minimal return path capabilities make applications such as video-phones difficult.

■ The short ranges from the base station require a relatively large number of base stations.

3.6 Summary of access technologies

Table 3.3 summarizes the access technologies discussed in this chapter.

Part II examines in more detail the environments and economics of WLL and shows why predictions for WLL networks are currently highly optimistic.

Table 3.3
Comparison of the Different Access Technologies

Access Technology	Data Rates	Advantages	Disadvantages
Voiceband modems	< 56 Kbps	Low cost; immediate installation	Blocks telephone line; relatively low data rates
ISDN	< 144 Kbps	Proven technology; relatively cheap	Only small improvement over voiceband modem; may be outdated rapidly
xDSL	~ 8 Mbps downstream, around 100 Kbps return	High data rate on existing lines	Unproven; expensive; will not work on all lines
Cable modems	30 Mbps downstream, 10 Mbps return	Relatively cheap; allows convergence	Has only limited penetration; architecture limits simultaneous users
TV distribution	Unknown, perhaps 10 Mbps downstream	Download of large data volumes to multiple subscribers	No return path; difficult to address individual homes
Mobile radio	64 Kbps cellular, 500 Kbps cordless	Can be used immediately where coverage	Limited data capabilities; cost; lack of coverage
Telephony WLL	< 384 Kbps	Economic to provide; reasonable data rates	High data rates not possible; new infrastructure required
Video WLL	500 Mbps downstream, 20 Kbps return	High data rates at low cost	Minimal return path; short range

References

[1] Webb, W., and L. Hanzo, *Modern Quadrature Amplitude Modulation*, New York: Wiley & Sons, 1994.

[2] Bocker, P., *The Integrated Services Digital Network: Concepts, Methods, Systems*, Springer-Verlag, 1988.

[3] Griffiths, J., ed., *ISDN Explained: World-Wide Applications and Network Technology*, New York: Wiley & Sons, 1990.

[4] Gillespie, A., *Access Networks*, Norwood, MA: Artech House, 1997.

[5] Kyees, P., et al., "ADSL: A New Twisted Pair Access to the Information Highway," *IEEE Commun.*, April 1995, pp. 52–60.

[6] Kerpez, K., and K. Sistanizadeh, "High-Bit Rate Asymmetric Digital Communications Over Telephone Loops," *IEEE Trans. Commun.*, Vol. 43, No. 6, June 1995, pp. 2038–2049.

Part II

Why Wireless?

ART I INTRODUCED the key trends in the global information industry and the plethora of access methods able to meet those trends. WLL was mentioned only as one of a range of possible access techniques; in comparison to the other techniques, WLL may appear to be lacking in bandwidth and with few obvious advantages to offset that potential problem.

However, between 1994 and 1997, over 130 WLL networks have been launched around the world, with analysts predicting that by 2005 over half the lines being installed will be WLL. The chapters in Part II explain why, despite its shortcomings, WLL has attracted such significant interest. The answer is in the details. Not all users want 10 Mbps, and not all PTOs are interested in providing that rate. When moving from sweeping generalizations of access technologies on a worldwide basis to a segmented analysis of different environments, the need for WLL in certain environments becomes clear.

In Part II, Chapters 4 and 5 develop the background information provided in Part I to show exactly where and why WLL is appropriate. They do so first by considering the different telecommunications environments in the world and analyzing the need for WLL in each of those environments. The analysis is complemented by a short history of WLL, illustrating where its key market areas lie. Chapter 5 builds on the qualitative analysis of different markets by considering the general economics of WLL and demonstrating its superiority in a range of situations. The aim of Part II is to demonstrate clearly where WLL can be deployed advantageously.

4

Telecommunications Environments Worldwide

T HE WORLD telecommunications environment is a highly eclectic place. Most, if not all, readers of this book have a phone at home, a phone at work, and a mobile phone. They have access to fax machines, possibly both at home and at work. They have e-mail facilities and occasionally surf the Net. Their main problem is redirecting calls such that the calls reach them wherever they are. It might come as a surprise to the readers of this book that over half the world's population have never made a phone call. The world average teledensity (the number of telephone lines in the world divided by the world's population) is only 2%. Even in Europe, teledensity varies from 60% in Sweden to 2% in Albania. The provision of lines to those who currently do not have access to telecommunications represents a vast untapped market.

There is a strong link between teledensity and *gross domestic product* (GDP) per head, as can be seen in Figure 4.1.[1] Further, there is strong evidence that good telecommunications can promote economic growth. That has prompted many governments in countries with low teledensity to seek ways to increase teledensity dramatically in the hope that such an increase will fuel domestic growth. The ITU recommends that teledensity should be at least 20% to foster economic growth or at least ensure that growth is not hindered by lack of telecommunications. For the purposes of WLL, the world's telecommunications environments often are segmented into three types:

- Countries with virtually no telecommunications (e.g., much of Africa);

- Countries with a low penetration (below, say, 40%) and long waiting lists (e.g., Eastern Europe);

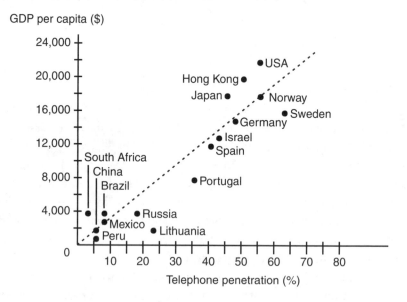

Figure 4.1 Link between GDP and teledensity.

---

1. It is not clear whether increasing GDP drives the provision of more lines or whether more lines help increase GDP. The answer probably lies somewhere between the two.

- Countries with high penetration (above, say, 40%), including countries where there is limited competition (e.g., France and Germany) and where there is high competition (e.g., the United Kingdom).

In regard to the potential for WLL, the three types of countries differ in a number of important ways:

- The functionality required by the users varies from simple telephony to high-speed multimedia.

- The ability of consumers to pay for the service varies from nonexistent to ample.

- The local telecommunications infrastructure varies from nonexistent to comprehensive.

- The competition in the access network varies from none to intense.

The differences are crucial to all aspects of the WLL network. As will become clear in Part IV, a key strength of WLL is its ability to adapt to each of those different environments through the selection of appropriate technologies and network plans in a manner no other access technology can emulate.

The remainder of this chapter considers each of the different environments in more detail before finally looking at how they have affected the history of WLL.

4.1 Developing countries

Any definition of developing countries is certain to be incomplete or to include some inappropriate entries, because of the breadth of countries that are described as such. That is not problematic in this chapter, which seeks to improve on the global generalizations of Chapter 3. It may, however, be a problem for a WLL operator considering a particular country, because the variations found in each country will ensure that off-the-shelf solutions are applicable only to a limited degree.

This section examines countries where the teledensity is below the ITU-recommended minimum of 20%. Another typical trait considered here is that there is a small or nonexistent waiting list for a telephone line because those without telephone access either cannot afford it or simply are unaware of the advantages it could bring. Countries included in this category are much of South America; most, if not all, of Africa; parts of the Middle East; India; much of China; remote parts of Russia; and some of the least developed countries in the Far East.

These countries typically have limited teledensity because the economics of the country are such that they cannot afford the cost of building a wired infrastructure. Much of the population live in self-contained communities needing little interaction with other communities; hence, the need for telephony is limited.

These countries may have great difficulties in installing telephony systems. With limited expertise in the country, expensive foreign workers are required to install and maintain the system. Climatic conditions such as tornadoes and flooding may regularly destroy any installed infrastructure. Even where climatic conditions are more favorable, the populations themselves may destroy the infrastructure through war or for commercial gain. In South Africa, for example, new copper cables stay buried only for a few weeks, after which the copper miraculously reappears on the local market in the form of bracelets and similar artifacts. Even fiber-optic cable seems to have some second-hand value.

Although the population may not be able to afford telephony, there may be forces at hand trying to introduce it. International aid agencies, such as the World Bank, recognize the pivotal role of telecommunications in leveraging a country out of third world status. Governments are equally aware of the importance of a good telecommunications system. Therefore, there may be funding for the installation and operation of the system even when the country apparently is too poor to afford it. As an example, again from South Africa, the government is using the revenue from cellular operators to finance a project to install a million phone lines over the next three years. Unfortunately, not all countries have the wealth from a minority in the populace to draw on in financing more universal telecommunications provision.

In such situations, clearly the key attribute of an access technology is low capital and through-life costs. Rapid installation also is important to avoid the need to borrow capital over long periods of time. High bandwidths and multiple lines per house would be unnecessary, and the complexity they entail needs to be avoided. The example of South Africa and the climatic problems mentioned previously demonstrate that cabled solutions simply are not viable in many of these countries. Wireless provides the only solution, offering the advantages of considerable through-life cost savings over twisted-pair installation, relatively low maintenance, and rapid rollout. Even the requirement for radio spectrum, often WLL's Achilles heel, as will be seen in later chapters, is only a minor difficulty in countries where much of the spectrum lies fallow in any case.

An example of such a country is India. Figure 4.2 shows the annual supply and waiting list for telephone lines in India in the last few years. It is clear that while the supply rate of lines in increasing significantly, it is having little effect on curbing the waiting list. Other interesting statistics concerning India include the following:

- Total area is 3.3 million km^2.

- Total population is 930 million.

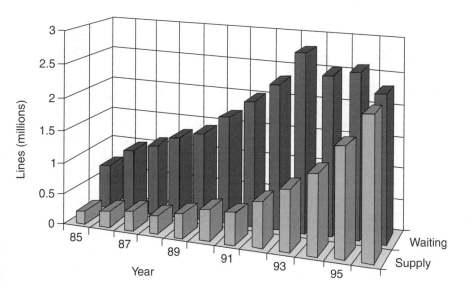

Figure 4.2 Supply of lines per year and average waiting list in India.

- Average teledensity is 1.45%.

- Rural teledensity is 0.2%.

- Number of villages without a telephone number is 400,000.

The Indian government has decided that the best way to increase the teledensity is to introduce competition to the incumbent PTO and to allow the use of WLL as the most economic and efficient means of supplying lines. A number of trials are underway, with demand forecast for the next three years as shown in Figure 4.3.

India, China, Africa, and South America contain over half the world's population. Projects underway in many of those countries to increase the teledensity represent a market for which WLL is the only contender. The key issue for the WLL operator will be whether the population can pay sufficiently high fees for the service to allow it to be profitable, taking into account possible subsidies and even, where the operator feels confident, the increase in wealth that an access system will bring.

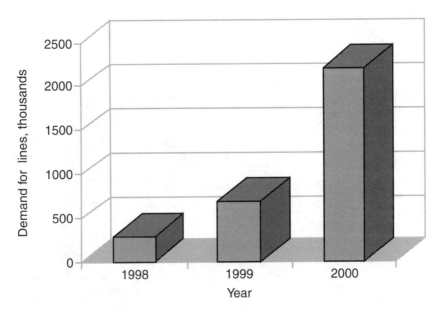

Figure 4.3   Projected WLL demand in India.

4.2 Eastern Europe

This category might be too specific because countries outside Eastern Europe also might fall into this category, while it may not apply to all Eastern European countries. Nevertheless, Eastern Europe generally epitomizes the countries in this category and represents a readily understood phenomenon.

This category includes countries with teledensities of between, say, 20% and 40%, but where the telephony environment is characterized by long waiting lists for telephone lines, often measured in years and even tens of years. In these countries, being able to afford the telecommunications service is not particularly problematic; the key issue simply is gaining access to it. Typically, the PTO has not been able to match demand, and no action has been taken to overcome that problem. This is the case in many communist countries where central planning failed to provide sufficient resources to developing telecommunications and where telecommunications often was viewed with suspicion by dictators who had concerns that it would permit subversive conversations. Many of these countries have moved to a more market-driven economy and recognize that rapid provision of telecommunications could significantly enhance their growth rates.

An example of a country in this category is Hungary. Hungary has 10.4 million inhabitants, a teledensity of 25%, and a mobile penetration of 6%. The government imposed obligations on the fixed-line providers that the number of subscribers should increase by a minimum of 15% annually. To do that, WLL systems were installed during 1995 and 1996 based on analog cellular technology. There are now 260 base stations serving 120,000 subscribers. The system has been judged a partial success, and much has been learned from it.

Other countries falling into this category include some of the Asia-Pacific countries, such as Indonesia and Malaysia, where recent high growth rates have resulted in a massive increase in demand for telecommunications, which the PTO has neither the resources nor the experience to meet.

In some cases, the demand has been met through the use of mobile communications, which has the advantage of rapid rollout and no waiting list. Such substitution, however, typically is expensive and only for those able to offset the costs against significantly enhanced business revenue. Mobile telephony helps ease the problem but does not provide an optimal solution for a country that requires an inexpensive, ubiquitous, and high-quality access network capable of future upgrade.

The key issue for all these countries is to rapidly provide telecommunications so that high growth rates can be achieved and waiting lists reduced to the months or days typical of first world countries. Currently most of the population would be happy with voice-only capability, but as a country rapidly gains in wealth, it seems likely that additional services such as limited data capabilities, Internet access, and similar services would be required. A forward-thinking operator would provision at least two 64-Kbps channels per user.

Unlike the developing countries, twisted pair and other cable solutions do provide possible alternatives in Eastern European–type countries. However, for whatever reasons, those solutions cannot be deployed with sufficient speed to meet the needs of the country. Solutions that provide more rapid connection include the following:

- Mobile radio, which typically is too expensive to meet the needs of everyone on the waiting lists;

- Satellite access, which has either no uplink access or is not suitable for deployment on millions of houses due to a lack of capacity and radio spectrum;

- WLL, which provides the most appropriate solution.

The suitability of WLL for these applications has been shown in numerous Eastern European countries. Table 4.1 lists some statistics relating to 1997 (these figures most likely will be significantly understated by the time this book is in print).

Eastern European countries will provide many WLL operators with environments the operators feel more comfortable with than they do in the developing world. There is a relatively stable and growing economy, a clear demand for the product, the means to pay for it, and sufficient expertise to help install and maintain the system. Competition is limited, with a single PTO (which largely has proved to be ineffectual at combating

Table 4.1
WLL Networks in Eastern Europe at the Beginning of 1997

Country	Number of WLL Networks	Technologies Adopted
Czech Republic	3	Various, including DECT and proprietary
Estonia	1	Analog cellular
Hungary	7	Mostly analog cellular, some DECT
Poland	4	CT-2
Romania	2	DECT and proprietary
Slovak Republic	1	DECT
Turkey	1	Proprietary
Uzbekistan	2	Proprietary

competition), few cable operators, and mobile operators charging high prices to the high-value market but not attacking the residential market. The requirements for services typically fall within the capability of WLL, assuring the operator that their network will not be quickly outdated.

Eastern Europe then represents an assured market. In comparison to the developed countries, the size of the market is small, but in absolute terms it probably exceeds 100 million lines, a substantial market by any estimate.

4.3 First world countries

By elimination, this category contains all the counties not included in the two preceding categories. To be more specific, this category includes countries with penetrations in excess of 40% and with waiting lists of less than one year. It includes Europe, North America, Canada, the Pacific Rim "Tigers," Australia, New Zealand, the rich Middle Eastern countries, and other similarly economically situated countries. The first world countries deliver the vast majority of the world's wealth and represent high-spending, high-value markets. In these countries, telecommunications operators have been able to match demand for the installation of new lines, typically have installed digital exchanges, and frequently increase their service offerings. California represents an interesting extreme, where demand has recently outstripped supply due to rapid growth in Internet access.

For the purposes of this analysis, it is instructive to split this category further into countries where the telecommunications environment has been liberalized and those where it has not. By "liberalized" is meant that the PTO is no longer protected from competition and that, with few restrictions, any operator wishing to provide telecommunications services may enter the market in competition with the PTO. One of the most liberalized environments is the United Kingdom, followed closely by the United States. Countries with limited liberalization include France and Germany, although to meet European Community directives, full liberalization will need to take place in these countries by 1998.

Both environments contain sophisticated consumers with multiple phone lines, high expectations, and increasing requirements for new services and increased bandwidth.

4.3.1 High penetration, limited competition

The countries in this category can be viewed as a progression of the Eastern European category. The PTO remains in a monopoly position, often heavily influenced by government and, as with many monopolies, typically is not as lean and mean as it might be. However, it at least has been able to meet demand, unlike the PTOs in the Eastern European category. Its service is disliked by many, and its charges are somewhat higher than could be achieved by a competing operator.

The main way that the government can change this situation is to introduce competition. Two forms of competition have been identified: competition in the trunk network and competition in the access network.

Competition in the trunk network often is the means whereby competition is introduced first. To compete in the trunk network, an operator merely has to interconnect key cities with high-bandwidth links. That can be achieved using buried cable, cables run along power lines, or cables sunk in places like canals and rivers. However, it does not take the PTO much to overcome such competition. By charging a high fee for access from the trunk network to the customers via the local network, a PTO can ensure that the effects of competition are minimal. Even despite strict regulation in the United Kingdom, trunk network competition from Mercury to BT failed to make any significant difference in prices and services. Trunk network competition, then, is unlikely to achieve governmental aims.

Competition in the local loop can provide significant stimulus to the PTO. It allows competitors to remove customers entirely from the PTO, a prospect that will stimulate most PTOs or simply result in competitors being able to take over the majority of their marketplace.

The government is likely to be eager to introduce competition into the local loop, and that represents an opportunity for the WLL operator. With only the inefficient PTO for competition, attracting consumers should be relatively easy.

Competition could come from a second wireline network or a wireless network. For the government, a wireless network is more attractive. New wireline networks cause massive disruption as streets are dug up, traffic disrupted, and environmental damage caused (e.g., the cutting of tree roots). Wireline networks take many tens of years to develop for the whole country, so the competitive pressures are slow to emerge. Governments, therefore, will want to encourage WLL operators by providing spectrum and licenses on favorable terms and generally helping when and where possible.

Although the government may be keen to encourage WLL operators, there is one key concern for the prospective operator, namely, that the competitive pressure will result in significant improvements to the PTO. That will allow the PTO to reduce prices and increase services, attracting customers back from the wireless operator to the PTO. The PTO may be able to introduce ISDN or xDSL services where needed, thus providing a better service than the WLL operator can. The WLL operator will want to recoup its investment costs over a relatively short period to reduce the risk that its network may become obsolete in the medium term. Given the high spending on telecommunications in these countries, that may be possible.

An example of such a market is Germany, where the government is only just starting to introduce competition. In Germany, the main threat to the PTO, Deutsche Telekom, comes from alliances between German industrial giants and outside telecom operators, the operators often having experience of competing in liberalized environments. Although Germany has a relatively high cable penetration, most of the cable network is owned by Deutsche Telekom; hence, there will not be competition between cable and twisted-pair operators in the same manner there is in other countries (e.g., the United Kingdom). The German trunked network market already is relatively competitive, so the government is concentrating on the provision of competition in the access network.

The first step toward such competition was taken in 1995, when the German government said it would grant 28 regional licenses to operate WLL networks using DECT technology. Governments generally get it wrong when they specify technologies, and this was no exception. Although DECT had the advantage that the frequencies were already available, and hence no long migration of the existing users of the radio frequency band was required, prospective operators were concerned that DECT might not be able to provide sufficient capacity, especially where there was more than one operator in the same area. Some of the higher capacity proprietary technologies were preferred. The government is now in the process of reconsidering how to issue the licenses, but the net result has been a significant delay in the introduction of competition.

There also still is concern over the regulatory environment. New operators are concerned that the regulator will not be sufficiently aggressive to prevent anticompetitive behavior by Deutsche Telekom. Already, Deutsche Telekom has provided discounts to some businesses that appear to be subsidized by other customers and designed to prevent "cream-skimming" by the new operators.

Overall, this represents a fertile area for new operators with only one main competitor. However, strategies need to be designed that will not allow cross-subsidization to work. For example, if the majority of residential subscribers are targeted, there will be nowhere for Deutsche Telekom to look for subsidy.

4.3.2 High penetration, high competition

In some countries, the telecommunications environment was liberalized some time ago and the competition for telecommunications has become intense. A good example is the United Kingdom, where there is competition in the trunk routes from BT, Mercury, Energis (which uses electricity pylons to carry fiber-optic cables), and other emergent operators. In the local loop, there is already competition between BT and over 150 cable franchises. In total, eight WLL licenses have been issued. There also is competition from mobile operators, some of which offer unlimited free calls in the evening and on weekends in an attempt to persuade customers to use mobile rather than fixed calls.

Competing in such an environment is extremely difficult. The PTO has responded to commercial pressure and is likely to continue both reducing call costs (by around 7% per year in the United Kingdom) and introducing new services such as call-back-when-free, voicemail, and caller ID. Moving into direct competition with the PTO is possible, but it is complicated by the need to continually reduce costs and increase functionality. The WLL operator also is likely to offer a relatively low data rate compared to the cable operators, ISDN lines, and, in the next five years, xDSL access technology.

Potential WLL operators must find particular niches in the market that they can penetrate. For example, in the United Kingdom, ISDN connection charges and annual rental fees are relatively high. It would be possible to offer ISDN to small businesses and individual homes that require Internet access at a lower price and hence gain substantial business. However, such a business move is predicated on the PTO not lowering its ISDN price to match those of the WLL operator, a likely response for a PTO that sees an important part of its market migrating to a different operator.

An alternative approach might be to couple telephony with broadcast through the use of MVDS, thus offering a differentiated proposition to the PTO (although not to the cable operator). By concentrating on areas where cable is uneconomic because the density of homes is too low, a potential market niche that cannot be attacked might be discovered.

Whatever approach is adopted in markets in the first world, the WLL operator must remember the key advantages and disadvantages. Advantages include the following:

- WLL is low cost relative to deploying twisted pair or cable (note, however, that the PTO will have depreciated the deployment cost many years previously).

- It offers high-speed deployment compared to twisted pair or cable, allowing customers to be attracted before the other operators can offer them service (once attracted, customers hopefully can be retained through loyalty schemes).

The disadvantages include the following:

- WLL cannot provide the access speed of other media and hence may be outdone in the services it can offer.

- The WLL operator cannot subsidize local loop access from other activities (such as international calls) in the same manner the PTO can.

In the market of high penetration and high competition, there is potential for WLL, but competitive pressures may be intense compared to the other categories. This market is illustrated next with a detailed look at the United Kingdom.

4.3.3　The U.K. marketplace

The United Kingdom is one of the most open and competitive markets in the world. Many other countries look to the United Kingdom for leadership on policies relating to telecommunications, and some large companies compete in the U.K. market, not primarily to make profit but to cut their teeth in a liberalized environment prior to entering the less competitive markets of the other European countries.

The first step toward liberalizing the United Kingdom market took place in 1984, when Mercury was licensed to provide public telecommunications services in competition with BT. The duopoly was broadly unsuccessful, however, in that Mercury did not make significant inroads into the BT market share. In 1991, the market was opened to further competition when cable operators were licensed to provide telephony over their cable networks. That went some way to improve consumer choice in areas where there was cable telephony, but the build rate remained slow. In 1995 the first WLL licenses were awarded in an effort to speed the provision of competition. Despite all the competition, BT has proved remarkably able at retaining a high market share. Privatization in 1993 helped in allowing BT to streamline its operation.

Since 1995, the U.K. government has licensed eight WLL operators. The first to be awarded licenses in 1995 were Ionica, in the frequency band 3.4 to 3.6 GHz, and Liberty, a Millicom International subsidiary, in the band 3.6 to 4.2 GHz. Ionica is now well into network rollout, having covered over 100,000 homes and gained around 15,000 subscribers by the start of 1997. Liberty's plans are not yet disclosed. The next licenses to be awarded were in 1996, when two licenses for rural use were offered. Unlike the previous licenses, BT was allowed to bid for one of those licenses because it was recognized that the use of those frequencies could help significantly in easing the *universal service obligation* (USO) on BT. (The USO states that BT must provide a telephone service for anyone in the country who requests it and is able to pay the fees. Furthermore, the fees must be the same for all users. In effect, the USO subsidizes the cost of telephony in rural areas from the revenues received in the urban area.) The licenses were duly awarded to BT and Radiotel. In a separate development, Atlantic was awarded a license for use in the Glasgow area in the 2.5-GHz band.

In the most recent awards of licenses at 10 GHz, Mercury, NTL, and Ionica have been awarded licenses to provide a service to businesses offering high bandwidths. Hence, given that there are more than 150 cable franchises, the total number of telecommunications operators in the United Kingdom can be seen to be much higher than most other countries.

It is still too early to understand the effect of the WLL operators on the U.K. market, because to date only Ionica and Atlantic have marketed a service. Both report that their early deployments have been successful, with Ionica reporting 3% penetration for East Anglia, the first region where they have rolled out. This level of penetration represents their break-even level, and they are confident that the network can operate profitably.

There have been recent changes in the regulatory philosophy imposed by the U.K. regulator of telecommunications, Oftel. From the time that BT was privatized, in 1993, until 1996, Oftel relied heavily on a regime in which BT's prices were forced to fall in real terms each year (the so-called RPI-x regime, where RPI stands for retail price index). Oftel argues that that regulation was one of the key reasons why BT's prices fell by 40% between 1993 and 1997, in real terms. New entrants disliked such regulation because they typically were forced to undercut BT to entice users to change to a different service. The RPI-x formula also affected the prices they could charge, making their competitive environment difficult.

Because of the fact that the regulatory regime did not significantly reduce BT's share and because of competitors' dislike of it, Oftel recently changed its regulatory stance to one in which prices are no longer fixed. However, Oftel increased its powers to deal with anticompetitive behavior.

Analysts have questioned whether BT's market share is really under threat from most of the new operators. Some new operators prefer to cream-skim by competing only for the highest value customers in the easiest-to-serve locations and thus maximize their profits. That implies that BT would retain the bulk of the market but lose the most profitable customers. Cream-skimming cannot continue indefinitely, however. As increasingly more WLL operators compete for market share, competition in the residential marketplace can be expected to become more intense. Figure 4.4 shows predictions from SBC Warburg of the BT market share in the coming years.

It remains to be seen whether the WLL operators will be more effective at reducing the BT market share than the cable operators have been. The general view is that they will achieve a greater success due to their significantly cheaper access technology and the fact that they can move much faster than the wired operators. Nevertheless, BT has resisted competition from Mercury and the cable operators and will fight hard to retain its market dominance.

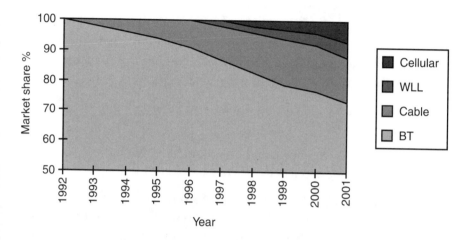

Figure 4.4 Predictions of market shares in the United Kingdom.

4.4 History of WLL

Wireless access first started to become a possibility in the 1950s and 1960s as simple radio technology at reduced price. For some remote communities in isolated parts of the country, the most effective manner of providing communications was to provide a single radio, kept in a central part of the community, that could be used to contact the operator and make a connection. That technology improved in performance and reduced in price during the 1970s, but also during that period copper networks were extended so the number of communities requiring radio access fell. By the end of the 1970s, communities linked by radio often had dedicated radio links to each remote house, the links connected into the switch such that they could be used in the same manner as normal twisted-pair links.

The next boost to wireless access occurred as fixed-link technology further improved and reduced in price, due in part to its widespread deployment for linking cellular base stations into switching sites. With the cost reduction, it became economic to link medium-size businesses using radio, even where access via twisted pair was straightforward. Competition often hastened this use of radio. For example, in the United

Kingdom, the second operator, Mercury, started using fixed links in the 13- and 22-GHz bands to interconnect businesses directly to their trunk network and thus avoid having to pay BT access charges. Such a use of fixed links gave Mercury the following advantages:

- Cheaper access. Cable did not have to be installed to the business customer.

- Faster access. Service could be provided within days of when the business requested it.

- Few sunken costs. If a business decided to move back to the PTO, Mercury could relocate the radio equipment to a different business; it would not have been able to relocate buried cable.

Such access, using point-to-point microwaves links, continues to be widespread to date. Operators using the technique still are recording growth in the number of links deployed.

WLL as a concept was established many years ago, if not recognized as an entire access network in its own right. That changed with the reintegration of East Germany and West Germany into a single unified Germany. Germany was not the only driver for WLL, and WLL certainly would have happened without unification. Indeed, there were developments prior to German reunification in which WLL networks were planned or deployed, but from the point of view of understanding WLL, Germany remains a good example.

When reunification of East and West Germany occurred, there was pressure to increase the standard of living in East Germany to that of West Germany in the shortest time possible. A key difference between the two parts of the country was teledensity, which was first world in West Germany and Eastern European in the East Germany. West Germany could provide the funding and the expertise to install twisted-pair access throughout East Germany, but that would be a slow process. In the interim, cellular radio seemed to offer a stop-gap measure to provide rapid telecommunications capability. So in East Germany, a number of cellular networks, based on the analog *Nordic Mobile Telephone* (NMT) standard, were deployed in the 800-MHz frequency band. The key difference was that subscribers had fixed units mounted to the sides of their houses to increase the signal strength and hence allow the networks to be constructed with larger cells and for lower costs. Thus, the first WLL network, as they are known today, was born.

From that point, the world of WLL grew at an incredible pace. Manufacturers of cellular radio and cordless radio technologies, quickly recognizing an additional market for their products over and above their initial design for mobile access, were quick to rename slightly modified products as "wireless local loop ready." The availability of equipment spurred governments into providing licenses and emerging operators to apply for them. In the first world countries, cellular systems were not appropriate because the spectrum for their operation was already in use, so manufacturers quickly started working on proprietary technologies. The number of announced WLL networks, which included trials and networks under construction, leapt from a handful in 1994, to double digits in 1995, up to around 50 in 1996, and by early 1997 had reached 130. The number of different technologies offered and different manufacturers providing equipment grew commensurably.

Compared to cellular, WLL still is in its infancy. Few, if any, networks are actually profitable yet. No consensus has been reached on the most appropriate technology, means of deployment, or countries suitable for WLL. Confusion continues to grow as microwave distribution systems are added to the wide range of WLL products. The history of WLL will continue to be written over the coming decade.

An interesting history of WLL developments from 1950 to 1990 can be found in a paper written by Claude Buster, the Chief of the Transmission Branch of the U.S. *Rural Electrification Administration* (REA). A summary of Buster's historical version of WLL, which has been reprinted in [1], follows.

- Early 1950s. Single-channel VHF subscriber equipment was purchased from Motorola, but the maintenance costs were too high as a result of the valve technology used and the power consumption too high. The trial was discontinued and the subscribers were connected by wire.

- Mid-1950s. Raytheon was given seed funds to develop 6-GHz band equipment, which would have a better reliability and a lower power consumption. The designers failed to achieve those goals, and the system still proved too expensive.

■ Late 1950s. Some equipment capable of providing mobile service to rural communities was put on trial. Users were prepared to pay a premium for mobile use, but the system still proved too expensive in a fixed application for which users were not prepared to pay a premium.

■ Early 1960s. Systems able to operate on a number of radio channels were developed, eliminating the need for each user to share a specific channel and thus increasing capacity. The general lack of channels and high cost, however, made these systems unattractive.

■ Early 1970s. A Canadian manufacturer developed equipment operating at 150 MHz that proved successful in serving fixed subscribers on an island in Lake Superior. The lack of frequencies in the band, however, precluded its widespread use.

■ Late 1970s. The radio equipment from several U.S. manufacturers was linked to provide service to isolated Puerto Rican villages. The service was possible only because the geographical location allowed the use of additional channels, providing greater capacity than would have been possible elsewhere.

■ Early 1980s. Communications satellites were examined for rural applications but were rejected as being too expensive.

■ 1985. Trials of a point-to-multipoint radio system using digital modulation promised sufficient capacity and reliability to make WLL look promising.

Indeed, the book by Calhoun is itself an interesting reference [1]. Calhoun used the book, written in 1991, as a vehicle to state his view that access methods would change from copper to radio in the future. Hence, even in 1991, it was far from universally accepted that WLL would become the predominant access method. Calhoun generally focused on the use of radio as a mechanism for rural access but looked further into the future to postulate that WLL might be the best way to provide ISDN access to fixed subscribers. Calhoun typically concentrated on the PTO switching from copper to wireless, and already the competitive environment has changed significantly.

4.5 The business of subsidization

Some customers cost more to interconnect to the network than others, yet all customers pay the same rental and call charges. The issue of subsidization is important for two reasons: first, the social implications of limited access into the telephone network and, second, the market opportunities it opens to incumbents. This section examines those issues in more detail.

Chapter 5 shows that the cost of connecting a subscriber with copper cable is almost entirely associated with the cable itself; the switching and interconnection costs are virtually insignificant. In principle, users closest to the local exchange should pay a lower fee for their calls than users farther away. With the exception of communities so isolated that satellite links are required, such differential costing never has occurred. That decision was made by both the PTO and the governments in their mutual interests. Governments always have been keen to achieve universal service provision, at least as far as possible, without having to fund the PTO explicitly. In return, the PTOs have been keen to retain their monopolies. PTOs have pointed out to governments that if they lost a monopoly over the customers who subsidize rural access, either they would no longer be able to afford to supply rural access or they would need rural bills to be an order of magnitude higher than urban bills. Neither of those options suited the government, so the status quo was preserved for many years.

The situation started to change when private operators were allowed to run their own microwave links. From there, it was a small step to ask why, once those links were installed, they could not offer capacity to others. In a number of landmark cases, courts ruled that such provision was legal, and in the 1970s competition was opened up in the U.S. backbone network, removing a key source of subsidization from the PTO. No longer was the PTO able to justify charging high rates for long distance calls when the additional resources required for the backhaul link cost virtually nothing. If the PTO did try to charge higher rates, it lost the business to competitors. However, subsidization in the local loop remains to this day. By analogy with the backhaul, it will be the imminent introduction of wireless local loop systems provided by competitors that will make such subsidization difficult to prolong. For more discussion of this topic, see [1].

The reason subsidization will become unsustainable is because competitors will be able to offer service to those who are doing the subsidizing at a much lower cost than they are currently paying. The PTO then loses the revenue it requires for subsidization, and the rural prices need to rise to overcome the loss of revenue. That is exactly what the cable companies have been doing in the United Kingdom, and it is likely to be a strategy adopted by WLL operators.

One important effect of WLL is that it reduces the relative cost difference between rural and urban provision. So, while rural access still remains more costly, the subsidization required is not so great. Even so, some WLL operators are actively targeting urban customers in preference to rural customers.

So what will be the effect of all these upheavals? It seems inevitable that subsidization gradually will be removed. Even if the PTO does not explicitly have different rates for urban and rural areas, those in urban areas will have the option of switching to a different telecommunications provider offering a cheaper tariff, and, in effect, rural customers will pay a higher rate. The societal implications concerning universal access will exercise politicians and courts in the coming years.

Chapter 5 looks at the fundamental economics underlying WLL systems. Previous chapters stated that an advantage of WLL systems is that they can be deployed more economically than wired systems. The next chapter explains why that is the case and shows how the cost advantage can be calculated.

Reference

[1] Calhoun, C., *Wireless Access and the Local Telephone Network*, Norwood, MA: Artech House, 1992.

5

The Economics of Wireless
Versus Fixed

ELECOMMUNICATION SYSTEMS, like many forms of business, require
a significant upfront investment followed by a period over which the
investment can be recouped in the form of revenue from customers.
The basic economics of a telecommunications system look somewhat
like this.

- An initial capital outflow to build a sufficiently large part of the
  network to make it viable;

- Interest payments on the capital sum until such period as the
  network becomes profitable;

- Revenue from customers increasing as more customers are signed to the network;

- Ongoing maintenance, upgrade, and marketing costs.

Fundamentally, over the investment period, which typically could be anything from 5 to 15 years, depending on the risks the operator is prepared to take, the cumulative revenue from the customers must offset all the costs associated with the network.

In terms of that economic model, the greatest difference between the different access technologies discussed in Chapter 3 is the initial capital cost, which naturally affects the interest payments. For the different access technologies, the revenue from the customers will be broadly similar, unless a particular access technology is able to offer some additional service (such as TV) for which the consumers will be prepared to pay additional fees. Many of the ongoing costs will be the same for all the different access technologies, particularly marketing, billing, and general customer-care issues. Some access methods, however, will have lower maintenance costs than others.

The key concern of WLL operators will be their cost base in comparison to operators using twisted-pair or cable access. Alternatively, WLL operators may be interested in whether they should install cable or themselves use wire. This chapter explains the key economic variables for wired and wireline access.

This chapter is concerned only with the access costs. All networks require backhaul connections and switches, which will be broadly similar for all access technologies. For that reason, those costs are not shown in detail here. Simplistically, a core network with switching and transmission equipment able to handle half a million users typically will cost around $15 to $30 million, a cost per user of $30 to $60. As will be seen, that cost is relatively insignificant compared to the cost of the access component.

Once the total network cost has been calculated, it must be factored into a complete business case, considering the borrowing requirements, costs relative to the competitors, sensitivity to change of key variables, and so on. Chapter 16 considers in detail the development of a business case.

5.1 The cost of wired systems

To provide wired access, it is necessary to provide a cable from a central distribution point, perhaps in the center of a city, to each house. To do that, it is necessary to dig up the road, lay cable, and then repair the road. Alternatively, particularly in rural areas, telegraph poles must be installed and wire strung between them. The cost of such an operation depends heavily on the environment: laying cable in a metropolitan area is considerably more difficult than laying cable across an open field. The costs also depend heavily on the local labor costs and productivity. As a guide, laying a cable in a city is likely to cost around $30 to $40/m, while laying a cable across a field might cost around $15 to $25/m. The cost to connect a house is given by its distance from the last house connected multiplied by the cost per meter of laying the cable. There also may be the cost of installing a suitable termination in the home. In the case of simple telephony, such cost is minimal. On the other hand, decoding boxes for cable TV typically cost around $300 to $400 each, depending on the functionality required. That cost will be borne by the network in some way, either in subsidy or in the need to charge lower prices so the total cost faced by the consumer falls below the cost of using a competitor.

The key variables in determining the cost are as follows:

- The distance between houses;

- Penetration (i.e., the percentage of homes passed by the cable that decide to pay for its use);

- The environment through which the cable must be routed.

For example, suppose that houses in a particular area are 10m apart and have front yards that are 5m long. The cost of cabling in the street is $30/m, while in gardens it is $20/m. Further, suppose that termination costs are $50 per home. The cost per home in the case of 100% penetration, then, would be as follows: $(10\text{m} \times \$30/\text{m} = \$300) + (5\text{m} \times \$20/\text{m} = \$100) + \$50 = \$450$.

Now imagine that the penetration drops to 20% (a figure typically achieved by cable operators), so it is necessary to pass four other houses before installing cable in a house. The cost to get the cable to the road outside a house is now 5 × (10m × $30/m) = $1,500. The other costs remain the same, resulting in a total cost of $1,650 per house.

Figure 5.1 illustrates the variations of cost per house based on distance between houses and penetration.

Because most PTOs installed their twisted pair at a time when there was near-100% penetration, then their costs will have been toward the lower end of those in Figure 5.1. Furthermore, the PTOs by now may have largely depreciated their installation costs. Hence, when a WLL operator is competing against a PTO, the difference in access costs between the two operators may be mostly academic. Instead, in this case, the comparison is between the maintenance costs of the wired system and the combination of the installation and maintenance costs of the wireless system.

Maintenance costs of buried cables are surprisingly high. Although the cables are underground, they suffer from flooding and from disruption caused by other entities digging up the streets, for example, to install and maintain gas and water mains. When problems occur, they often require that a stretch of street be redug to locate the fault and replace the cable. Maintenance costs on buried cable are typically some 5% of

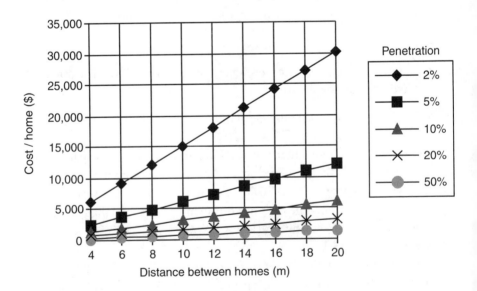

Figure 5.1 Variation of cost with penetration and distance.

the installation cost per year, whereas maintenance costs on a wireless system might be between 1% and 2.5% per year.[1]

As Calhoun points out, that approach to costing is simplistic [1]. It takes into account only the cost of the dedicated cabling required for a specific house. However, each house also shares part of the trunked network used to bring the cable up to the house. Further, there are many other variables such as the wiring configuration, the type of ground, the potential need to replace the entire backbone when its capacity is exceeded, and so on, which will complicate the issue. However, there seems little industry agreement on how to treat such costs in the numerous and diverse deployments that exist at the moment. That probably is a matter best left to a fixed network specialist. It is unlikely to change the overall flavor of the analysis presented here.

5.2 The cost of wireless systems

The economics of wireless systems are completely different from those of wired systems. To provide a wireless service, an operator must erect a transmitter. Service then is offered to the customers in the coverage area. Those who require the service are provided with a receiver unit to be mounted on the side of their house (as with cable, customers may pay for the unit, but its cost is borne in some fashion by the operator). If a subscriber subsequently no longer requires service, the subscriber unit is removed from the side of the house and installed elsewhere.

The key costs for WLL are the cost of installing a transmitter, the number of subscribers for whom the transmitter provides service, and the cost of a subscriber unit. Those costs vary dramatically across the different WLL technologies, as will be explained in detail in Part IV. For the moment, assume that a transmitter costs $150,000 to buy and install, that it covers an area with a radius of 4 km, and that subscriber units cost

---

1. Until WLL systems have been in deployment for a number of years, the typical maintenance cost will not be known. Perhaps the best comparison is with cellular systems, which typically have a 1% maintenance cost. However, WLL networks also need to maintain subscriber equipment, which may push the cost higher.

around $400 each. The key variables determining the cost, then, are the following:

- The number of houses in the coverage area of one transmitter;
- The costs of the subscriber units.

Suppose, for example, that there are 1,000 houses in a cell with a radius of 4 km, that the transmitter costs $150,000, and that subscriber units are $400 each, including installation. The fraction of the transmitter cost applying to each house, then, is $150, and the total cost per house is $550. That is slightly more than the $450 per house calculated in the cable case.

Now suppose that penetration is only 20%. The transmitter cost per subscriber rises to $750 per house and the total cost to $1,150. That is significantly less expensive than the $1,650 calculated in the cable case.

The graphs in Figures 5.2, 5.3, and 5.4 show the relative costs of cable access and wireless access for a range of housing density and penetration levels. The advantages of WLL as the housing density falls or the penetration falls are obvious.

The figures show WLL consistently providing a less expensive access technology than cable, with the difference decreasing as the penetration increases. The costs of both systems rise as the density of homes falls, with WLL remaining the least expensive alternative.

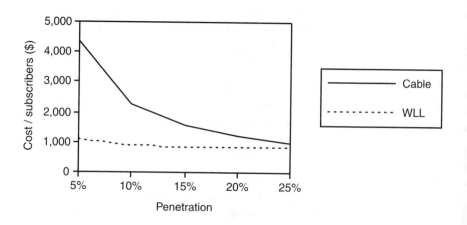

Figure 5.2 Relative costs of cable and WLL in a high-density case.

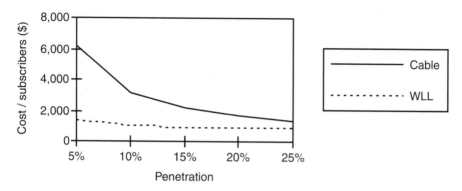

Figure 5.3 Relative costs of cable and WLL in a medium-density case.

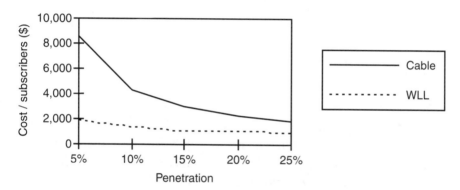

Figure 5.4 Relative costs of cable and WLL in a low-density case.

A new operator faced with the choice of using wired or wireless access techniques probably should adopt a wireless approach. Of interest to operators in first world countries will be whether they can compete against the incumbents, particularly where the incumbent already has 100% penetration and has depreciated installation costs. In that case, a comparison is made between the maintenance costs of wired and the installation and maintenance costs of wireless. Taking an average maintenance figure of 1.25% for wireless and 5% for wired and using the same assumptions as in the earlier example, Figure 5.5 shows the relative costs of an incumbent using depreciated wire and of a new entrant using a WLL solution, considered over a 15-year period and taking into account the cost of capital.

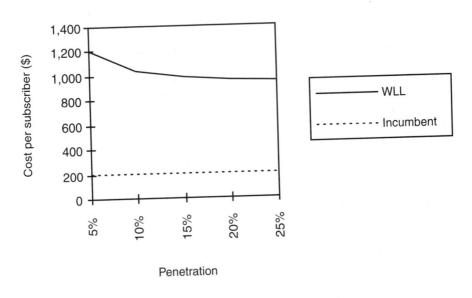

Figure 5.5   Through-life costs for incumbent and new operator.

That presents quite a different picture. Although its maintenance costs are higher, the incumbent no longer has to bear the installation costs. According to Figure 5.5, it appears that the new operator should not attempt to compete with the incumbent. However, there are a number of reasons why the incumbent's costs may be higher. First, the incumbent, as a previous monopoly supplier, is likely to be somewhat inefficient and hence to have a higher overall cost base than is suggested here. Second, the incumbent is likely to have a USO; as discussed in Section 4.5, that typically means that the incumbent subsidizes loss-making customers through other customers, increasing the cost per customer shown in Figure 5.5. Third, the wireless operator can embark on cream-skimming, in which they cover only those areas where the economics are particularly favorable.

Competing with an incumbent certainly is possible, for example, in the United Kingdom, Ionica is successfully gaining market share and is predicting profitable operation, despite apparently higher costs. However, it requires considerably more skill than the case in which the incumbent is nonexistent or is providing an inadequate service.

5.3 Hybrid systems

So far, this chapter has examined two disparate alternatives. Many observers expect that wired and wireless access techniques will exist side by side in the same network. One possibility is the so-called doughnut cell. In that scenario, the cost of connecting homes very close to the switch might be lower by using copper rather than radio because of the short distances involved. However, above a radius of, say, 1km, the costs are lower by radio. Thus, for a base station placed at the switch, the wireless subscribers exist in a doughnut-shaped ring.

Of course, it is unlikely that existing operators will rapidly remove the copper and replace it with radio. Instead, as BT has done, radio is being used initially in rural areas and increasingly might be used in new housing developments or areas where the maintenance costs of copper have become unacceptably high.

The possibility of a hybrid approach adds an additional complexity to any costing analysis. As will be seen in later chapters, that complexity can be factored into an appropriate business modeling spreadsheet.

5.4 Market forecasts

Chapter 4 showed that in certain environments there are very strong reasons why WLL systems should be deployed. This chapter has shown, simplistically, that the economics of WLL can be highly favorable, particularly where the density of homes is low or the penetration achieved is low. Numerous manufacturers, potential operators, and market analysts have reached those same conclusions. This section looks at some recent forecasts for WLL. Forecasting in this area is extremely difficult, given the huge number of unknowns, such as whether some of the developing countries will be able to afford a telecommunications system. Therefore, a large variation in the forecasts is to be expected. The forecasts given here were made during the latter part of 1996 and the early part of 1997; as time progresses, it is likely that the differences between the forecasts will narrow.

Figure 5.6 shows some recent forecasts as to the size of the market in terms of number of installed lines in the year 2000.

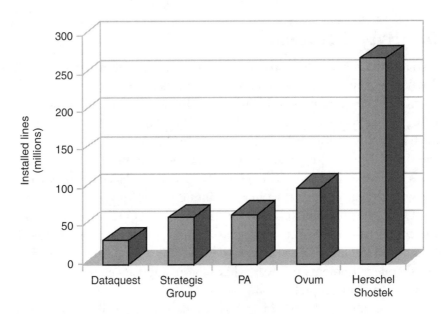

Figure 5.6   Market predictions for WLL made at the start of 1997.

PA predicts that by the year 2000 over 10% of all lines being installed will be wireless, while others have predicted that the number of wireless lines installed per year could overtake wired lines before the year 2005.

The market forecasts for the year 2000 give an average of more than 100 million lines. With a cost per line perhaps in the region of $1,000, that represents a market of approximately $100 billion by the year 2000 and growing substantially thereafter. Whatever happens, WLL is going to be big business.

Part II has shown why WLL is of interest; Parts III through V concentrate on explaining WLL systems themselves in considerably more detail. Part III will be useful for those readers who want a technical understanding of the different technologies. Those who are not of a technical bent can skip Part III.

Reference

[1]   Calhoun, C., *Wireless Access and the Local Telephone Network*, Norwood, MA: Artech House, 1992.

Part III

Technical Information About Wireless

ARTS I AND II introduced the general telecommunications scene and the potential role of WLL in that scene. The remainder of this book focuses on describing WLL systems in more detail.

WLL systems are radio systems, and to understand them fully it is necessary to know something about radio. This part of the book provides some relevant background on technical matters associated with radio propagation, radio systems, and radio access methods. Propagation is critical in reaching the subscriber reliably. Radio system design is critical in ensuring optimum performance for minimum cost, while the access method forms one of the key distinguishing factors between the different technologies. For that reason, it is useful to understand radio systems in detail.

Numerous books describe these technical subjects in great detail. This text provides detail sufficient only to understand the implications for WLL. For those wishing more detail, a particularly good reference that fits well with the topics presented here is *Wireless Information Networks*, by K. Pahlavan and A. Levesque (New York: Wiley & Sons, 1995).

6

Radio Propagation

RADIO PROPAGATION PLAYS A KEY ROLE in WLL systems. It is the means whereby signals are transmitted from both the transmitter and the subscriber unit. The propagation that can be achieved limits the range of the base station, requiring more cell sites than might otherwise be required. Propagation phenomena mean that some subscribers in a cell are unable to obtain a satisfactory signal. Propagation causes some frequencies to be more desirable than others. Understanding propagation is key to understanding WLL technology.

6.1 The environment of mobile radio propagation

Much work on propagation has been performed by those involved in cellular radio. Cellular propagation is complicated by the movement of the mobile, possibly into buildings and areas shadowed from the radio signal. WLL propagation is substantially less complicated than that. However, some of the phenomena that cause such difficulty to mobile radio propagation have an impact on WLL systems and are discussed in this section.

Before examining WLL in more detail, it is worth reviewing briefly the mobile radio propagation environment and the phenomena that affect it. Mobile radio planners consider that the path loss experienced when a signal is transmitted through the channel is composed of three distinct phenomena:

- Distance-related attenuation;

- Slow fading;

- Fast fading.

6.1.1 Distance-related attenuation

Distance-related attenuation simply expresses the fact that as the distance from the base station increases the signal strength decreases. That is entirely consistent with everyday experience: the farther one moves from someone who is talking, the weaker the signal. The drop in signal strength is caused by the fact that the signal spreads out from the source on the surface of a sphere. The area of the surface is proportional to the radius squared; hence, the signal strength is proportional to $1/d^2$, where d is the distance from the transmitter.

Measurements of mobile radio channels have found that in practice, the signal strength decreases more quickly than $1/d^2$. Typical values often used in predicting mobile radio propagation are $1/d^{3.5}$ or $1/d^4$, depending on the model used. The reasons for that more rapid reduction in signal are the following:

- The presence of the ground interferes with the expansion of a spherical surface, resulting in only a hemisphere. The conductivity and reflectivity of the ground then determine to what extent propagation is affected.

- Signals are attenuated by vegetation and buildings, and the loss associated with passing through or around those things tends to increase the propagation exponent.

In the case of WLL, most of the propagation is via direct *line-of-sight* (LOS), as explained in Section 6.2. In this case, the second of the two reasons for the exponent deviating from 2 no longer applies, and the first effect is extremely weak, with the result that a path-loss law of $1/d^2$ typically is experienced.

6.1.2 Slow fading

Slow fading is a mobile radio phenomenon caused by the mobile passing behind a building. During the period the mobile is behind the building, the signal received is reduced. Driving along a road, the mobile will pass behind a sequence of buildings, causing the signal to reduce in strength, or fade, on a relatively slow basis (compared to fast fading). This phenomenon is not directly applicable to WLL, because the receiver does not move. However, receivers installed where the path to the transmitter is shadowed are, in effect, in a permanent slow fade. Understanding the loss of signal due to such a fade allows planners to assess whether subscribers who are shadowed are able to receive sufficient signal. Such an assessment is based on an understanding of diffraction and reflection. If a signal is received via a reflection in a WLL system, it typically is about 15 dB weaker than if it is received via a direct path. If the use of reflections is needed in some areas, then that margin needs to be added into the path-loss budget.

6.1.3 Fast fading

Fast fading is another mobile radio phenomenon. It is caused by the signal arriving at the receiver via a number of paths. Imagine that two signals are received at the mobile. One passes directly from the base station to the mobile via a line-of-sight path. The other is reflected off a building behind the mobile and back into the mobile antenna, as shown in Figure 6.1. The mobile then sees a signal that is the composite of those two signals. The reflected signal has traveled a slightly longer path than the direct signal and thus is delayed slightly compared to the direct signal. The result of that delay is that the phase of the reflected signal differs from that of the transmitted signal. The phase difference is related to the difference in distance multiplied by the speed of light (giving the delay on the signal) multiplied by the frequency of transmission.

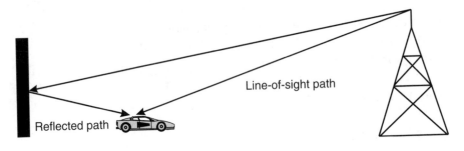

Figure 6.1 Multipath propagation.

Fast fading can be better understood by an example. Suppose the reflected signal travels an additional 10m before arriving at the mobile. Light travels at 3×10^8 m/s, so the additional distance causes a delay of around 30 ps (30×10^{-9} sec). If the frequency of transmission is 3 GHz, one cycle of the carrier wave at that frequency takes 0.3 ps. Thus, the reflected wave is delayed by 100 cycles of the carrier wave exactly. Every additional 10-cm distance, in fact, delays the reflected wave by a further cycle of the carrier wave.

When the reflected wave is delayed by an exact multiple of a cycle of the carrier wave, the two received signals are said to be in phase. The two signals sum, such that the total received signal has twice the strength of the direct signal. When the reflected wave is delayed by an exact multiple plus exactly a further half-cycle, the two signals are said to be in antiphase and cancel each other exactly. That is to say, at a certain point, the mobile loses all received signal. By moving a distance of only half a wavelength then, some 5 cm, the mobile moves from a position where the signal strength is doubled to one where there is no received signal. That phenomenon repeats continuously as the mobile moves, so the signal fades rapidly, giving the effect its name.

Of course, in real life, there are many more that just a single reflected signal path, and the reflected signals are not all equal strength, so the prospect of an exact cancellation is somewhat reduced. Nevertheless, fading is a severe problem. Figure 6.2 shows a typical fading waveform, often termed Rayleigh fading, after the mathematician who developed the statistics that can be used to describe such a waveform. It can be seen that in a period of 1 sec a number of fades, some as deep as 40 dB, are experienced. This book is not the place to discuss the mathematics of

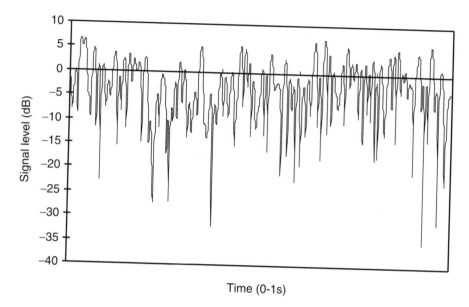

Figure 6.2 Rayleigh fading waveform for a mobile moving at walking speed over 1 sec.

Rayleigh fading; those interested in learning more can consult almost any book that discusses mobile radio propagation, for example, [1].

A further difficulty may be caused by fast fading. Imagine that instead of the reflection coming from a nearby building, it comes from some faraway mountain. The delay of the reflected signal may now be quite large. If the delay is greater than the time taken to transmit a bit of information, then when the reflected signal finally arrives, it is carrying different information to the direct signal. The result is that the previous bit transmitted, or symbol, interferes with the current symbol, generating a phenomenon known as *intersymbol interference* (ISI). The problem is akin to listening to a public address system when there are a number of loud speakers: the signal from the farthest loudspeakers is so delayed that a syllable arrives during the time when the next syllable is being heard from the nearest loud speaker, making comprehension difficult.

Fast fading is less relevant to WLL. Typically, where there is a LOS path, there is little fast fading because any reflected signals tend to be so much weaker than the LOS signal that the cancelation effect is minimal. Only where the main path is shadowed is fast fading likely to be a problem.

During installation of the receiver equipment, it is important to mount the equipment such that it is not in a fade (which tends to be more or less stationary in space); small movements of around 5 cm of the receiver on its mountings might be required to receive the strongest possible signal. If, during a call, a bus drives by, for example, further reflections off the bus may be generated. Those reflections will temporarily change the reflection pattern, possibly causing fading to occur, each fade happening as the bus moves around 10 cm. Hence, WLL equipment, especially that installed in shadowed areas, needs to be tolerant to fading occurring from time to time. The implications of this on system design are described in Chapter 7.

6.2 The line-of-sight channel

In mobile radio, a LOS between the mobile and the base station is a rare occurrence. WLL systems, however, typically need to be designed such that a LOS is more frequently achieved. That is because at the frequencies used for WLL, radio waves do not diffract well around obstacles. Hence, any obstacle tends to more or less block the signal resulting in inadequate signal strength for good reception. Further, WLL systems need to provide a voice quality comparable with fixed systems, unlike cellular, which is able to provide an inferior voice quality. To obtain a high quality, a low *bit error rate* (BER) channel is required. Such channels typically require a LOS path to obtain an adequate SNR.

This is not too problematic for WLL, because the receiver units can be placed on building roofs, where a LOS is much more likely than would be the case for mobile radio. Establishing whether there is a LOS channel is straightforward—if the location of the subscriber antenna is visible (by eye) from the transmitter site, there is a LOS. (In practice, visual surveys of potential transmitter sites are time consuming, so computer models are used to help predict LOS paths.)

Despite earlier comments concerning the need for a LOS channel, signals do diffract to a limited extent, and reflections can often provide an adequate signal level. Figures 6.3 and 6.4 are examples of diffracted and reflected paths, respectively.

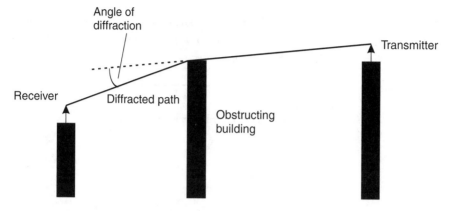

Figure 6.3 Example of a diffracted path.

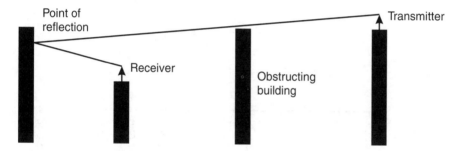

Figure 6.4 Example of a reflected path.

6.2.1 Diffraction

Diffraction is a phenomenon caused by the fact that electromagnetic waves propagate as if each point on a wavefront generates a new wave. As the wave grazes the top of a building, wavelets are emitted in all directions, including away from the LOS path. The farther away from the LOS path, the weaker the signal. The extent of diffraction depends on two parameters, the diffraction angle (i.e., the angle through which the path needs to "bend" as it grazes the top of the building obscuring the receiver to arrive at the receiver) and the frequency of the carrier wave. The greater the angle, the lower the received signal, and the higher the frequency the lower the received signal. Readers interested in a mathematical treatment of diffraction should consult texts on signal propagation. Figure 6.5 shows the variation of signal strength of a diffracted signal with parameter v, and Figure 6.6 shows how the signal strength for a given diffracted angle varies with frequency.

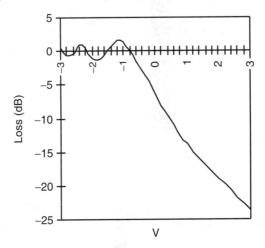

Figure 6.5 Variation of diffraction loss with parameter v.

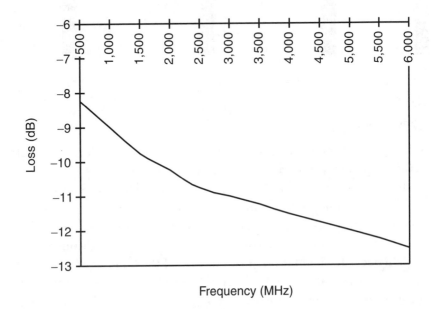

Figure 6.6 Variation of diffraction loss for a particular obstruction with frequency.

The parameter v is given by

$$v = h\sqrt{\frac{2(d_1 + d_2)}{\lambda \cdot d_1 d_2}} \qquad (6.1)$$

where d_1 is the distance from the transmitter to the obstruction, d_2 is the distance from the obstruction to the receiver, h is the height of the obstruction, and λ is the wavelength of the transmitted frequency. As a guide, for a possible WLL deployment with $d_1 = d_2 = 500$m, at 900 MHz the parameter $v = 0.16h$, while at 3 GHz, $v = 0.28h$. Using Figure 6.5, if $h = 0$ (i.e., the LOS path grazes the top of a building), the diffraction loss will be around 6 dB, while if the height is 5m (corresponding to an angle of 1.14 degrees), the loss will be 12.5 dB at 900 MHz and 16 dB at 3 GHz. Thus, diffraction angles greater than 1 degree are likely to result in insufficient signal strength at the frequencies of interest. As Figure 6.6 shows, the loss is frequency dependent, but because the loss already is severe, the frequency variation is unlikely to be the overriding issue.

6.2.2 Reflection

Reflection is caused when a wave strikes an object and is reflected back from it. It is, of course, the phenomenon by which we see the world around us. Different materials reflect to a different extent, the amount of absorption by the material being a key parameter in determining whether a sufficient signal strength will be achieved. In practice, concrete and glass buildings provide quite good reflections. Like light, the receiver needs to be in just the right place to see a reflected image of the transmitter (compare that with needing to be in the right place to see yourself in a mirror). Many building surfaces, however, are "rough," so the reflection is scattered on hitting the building. That makes the zone where the reflections can be received larger but the signal strength received weaker.

The key difficulty for WLL, with its directional antennas at the subscriber equipment, is finding the reflections. An installer that is unable to obtain adequate signal strength when pointing the antenna toward the base station may need to turn the antenna through up to 360 degrees in an attempt to find suitable reflected paths. In many cases those will not exist.

Slightly more complexity is associated with the LOS path than has been explained so far. Readers may have been perplexed to notice that even when the diffraction angle is 0 degrees (i.e., the LOS path grazed an obstruction), there still is a significant loss. That is caused by a phenomenon known as the Fresnel effect. According to Huygens-Fresnel theory, the electromagnetic field at a receiver is due to the summation of the fields caused by re-radiation from some small incremental areas over a closed surface about a point source at the transmitter. The field at a constant distance from the transmitter, that is, on a spherical surface, has the same phase over the entire surface. That surface is called a wavefront. If the distances from the wavefront to the receiver are considered, the contributions to the field at the receiver are seen to be made up of components that add vectorially in accordance with their relative phase difference. Any object blocking part of the wavefront affects the signal received.

It is possible to draw ellipsoids from the transmitter to the receiver, where the distances from the transmitter and from the receiver differ by multiples of a half wavelength. Each of those ellipsoids is known as a Fresnel zone. Broadly, if there is any obstruction in the first half of the first Fresnel zone, there is a significant reduction in signal strength. The radius of the first Fresnel zone can be approximated by

$$R = \sqrt{\frac{\lambda \cdot d_1 d_2}{d_1 + d_2}} \qquad (6.2)$$

where the terms are as defined in Subsection 6.2.1. To avoid any loss of signal through interference with the first Fresnel zone, the closest obstruction to a 1-km path must come no closer than 4.3m to the LOS path for a 1-GHz link. Readers who want to know more about Fresnel effects are referred to [2].

Using LOS, diffraction, and reflection, subscriber units can be installed in such a manner that they receive sufficient signal strength.

6.3 Time variation in channels

Unfortunately, once installed, the signal will not remain static. Three key effects will result in the signal received varying over time:

- Fast fading;
- New obstructions;
- Rainfall.

6.3.1 Fast fading

Fast fading was explained in Section 6.2. If there is not a LOS path, or there are particularly strong reflections, then when large objects move in the vicinity of the receiver the fading pattern will change. If the carrier frequency is 3 GHz, and a bus moves at 50 km/h, then fades will occur every 10 ms. In the worst case, the signal might decay momentarily to zero during a fade before being restored to full strength. There is little that a planning engineer can do to overcome that problem. Instead, the radio equipment designer must ensure that the equipment is able to correct the errors occurring from that phenomenon.

6.3.2 New obstructions

New obstructions occasionally arise. The most obvious example is the construction of a new building between the transmitter and the receiver, blocking a previous LOS path. Such blocking is often catastrophic in rendering the received signal too weak to be of use. Clearly, a radio planner or radio design engineer can do little about that problem. The extent of the problem will vary depending on the rate of build in the area. The WLL operator should be aware of new buildings under construction and model the effects to look for possible solutions, such as using reflections, building an additional cell or repeater, or aligning the receiver toward the base station in an adjacent cell.

6.3.3 Rainfall

Rainfall can be a problem. Radio signals are attenuated by moisture in the atmosphere. The level of attenuation varies with the carrier frequency, the quantity of rain falling, and the distance of the transmitter from the receiver. The variation of attenuation with frequency is particularly strong and highly nonlinear. At 3 GHz, the highest attenuation is around 0.06 dB/km; for a typical WLL path of, say, 6 km, the attenuation is only 0.36 dB. However, at 10 GHz, the attenuation could, in extreme cases, be as high as 5 dB/km, or up to 30 dB over a typical link. Such a loss of signal almost certainly would make the link unusable. Above around 5 GHz in countries where heavy rainfall is normal, rain fading is a potential problem.

Broadly speaking, the heavier the rain, the higher the attenuation. Countries that experience extremely heavy rainfall from time to time, such as those with monsoon climates, are most affected by this problem. Rain fade results in an increased attenuation per meter, so the farther away the receiver, the greater the problem. Receivers on the edge of cells are first to suffer during heavy rain.

Where rainfall attenuation is problematic, it must be allowed for in the link budget. Based on the frequency, the expected distribution of the rainfall and the outage time that can be tolerated (if any), the additional path loss that needs to be allowed for can be calculated and the resulting reduction in propagation range determined. Cells then need to be designed taking this into account.

6.4 Wideband channels

As mentioned in Subsection 6.1.3, reflection can cause ISI. That is problematic in mobile radio systems having transmission bandwidths greater than around 100 kHz. It is far less a problem with WLL. To understand whether the problem might occur, consider the following example:

> Assume that the symbol rate is 100 kHz (in a broad sense, that means the bandwidth of the transmitted signal is around 100 kHz). The carrier frequency is not a relevant factor in calculating ISI. (The symbol rate, for the moment, can be assumed to be equal to the bit rate; the situations where that is not true are discussed in Chapter 7.) If the reflection is delayed by 1/100 kHz, namely 10 µs, the reflected signal

arrives during the next symbol period causing ISI. In 10 μs, radio signals travel 3 km. Thus, if the reflection occurs from an obstacle 1.5 km past the receiver, ISI occurs.

As the symbol rate increases, the distance required for ISI to occur reduces proportionally. With a bandwidth of 1 MHz, the distance falls to 300m, and at 10 MHz the distance is down to only 30m. Some of the WLL systems proposed (as discussed in Part IV) have bandwidths in excess of 1 MHz; hence, ISI might be expected to be problematic.

Compared to mobile radio, WLL systems, even those with high bandwidths, tend to avoid ISI. That is because long reflected paths typically result in signals being reflected from behind the mobile back into the mobile. The directional antennas typically used in WLL mean that such reflected paths are highly attenuated by the radiation pattern of the antenna in the off-axis direction and thus fall below the receiver sensitivity threshold.

Further, without preempting the discussion on access technologies in the next two chapters, one particular technology known as *code division multiple access* (CDMA) is broadly immune to ISI. Many of the WLL technologies proposed with wide bandwidths utilize CDMA technology.

In summary, apart from a few exceptional cases, WLL radio planners need not be too concerned with ISI.

6.5 Frequencies for WLL

As has been mentioned, some propagation phenomena depend on the frequency of transmission. If this were a book on mobile radio, the variation with transmission frequency would be of little relevance, because across the world the same frequency bands are used for mobile radio. Unfortunately, that is not the case for WLL. To understand why, it is necessary to know a little more about the background to radio spectrum allocation.

Radio spectrum is a scarce resource; as one American analyst noted, "Spectrum is like real estate—they just don't make it any more." Numerous applications require radio spectrum, including cellular, microwave links, satellites, radars, meteorology systems, government radio systems, defense equipment, garage door openers, and microwave ovens. The list is almost endless. As a result, new systems need to find parts of the spectrum that are little used and fit in with existing users. Those little used parts tend to be at higher frequencies, where the propagation achieved is poor compared to lower frequencies.

More recently, the scarcity of spectrum at different frequency bands has been quantified as spectrum has been auctioned in countries such as the United States. The value under auction, of course, relates to the demand for spectrum, which in turn relates to the profit that can be made by the organization that owns the spectrum. Because users are prepared to pay a premium for mobile phones, spectrum generally is more valuable for mobile radio than WLL. WLL operators then are forced to use higher frequencies than mobile radio operators. Only in countries where mobile radio is not widely used can the WLL operator gain access to the cellular frequencies.

The need for low frequencies (i.e., below 1 GHz) with their resultant long propagation ranges also varies across different WLL deployments. Where the subscribers are distributed with a low density across a large area, long range is important. Where users are clustered in a city, a long range can actually be problematic, resulting in interference to neighboring cells. Because of those differences, different WLL operators have looked to different frequencies to best meet their needs.

Attempts are ongoing to standardize some frequencies across the world for WLL operation. Standardization would enable manufacturers to achieve economies of scale and operators to minimize interference with neighboring countries. Because around the world, the higher frequencies are lightly used, WLL frequency standardization must, by necessity, focus on frequencies above 3 GHz. The result is that operators requiring greater range than can be achieved at 3 GHz will use nonstandardized frequency bands. There is little problem with that, as long as an assignment can be gained in the country of operation and equipment obtained that operates at the required frequency.

It is not appropriate here to discuss the process of standardizing frequencies; readers who are interested should refer to [3]. Suffice it to say that at the time of this writing it seemed likely that before the end of 1998 3.4 to 3.6 GHz and 10.15 to 10.65 GHz would emerge as standard bands. As will be seen in Part IV, only a fraction of the currently available WLL systems are designed to work at those frequencies.

Table 6.1 lists the frequencies currently used or proposed for WLL. From the discussion in this chapter, readers will have deduced that the lower the frequency the greater the range and the better the diffraction around obstructions. Broadly speaking, then, operators tend to try to obtain spectrum at as low a frequency as possible. A number of factors tend to put pressure on operators to agree to using spectrum at higher frequencies, including the following:

- The higher cost of lower frequency spectrum where it is auctioned;

- The fact that more spectrum is available at higher frequency, providing a greater system capacity;

- The fact that the size of the cell may be limited not by the propagation distance but by the number of subscribers that can be supported in a cell; if there are more subscribers than the maximum in a cell, the size of the cell must be reduced.

Those factors are discussed in more detail in Part V. Suffice it to say here that global variations cause all these frequencies to be of use to some operators. Thus, the wide variety of frequencies in use can be expected to continue, and indeed increase, in the future.

Table 6.1
Frequencies Used or Proposed for WLL

Frequency	Use
400–500 MHz	For rural application
800–1,000 MHz	For cellular radio in most countries
1.5 GHz	Typically for satellites and fixed links
1.7–2 GHz	For cordless and cellular bands in most countries
2.5 GHz	Typically for industrial equipment
3.4–3.6 GHz	Being standardized for WLL around the world
10 GHz	Newly opened for WLL in some countries
28 GHz and 40 GHz	For microwave distribution systems around the world

6.6 Predicting WLL coverage

To minimize system cost and rollout time, WLL operators need to ensure that they use as few cell sites as possible to provide the required coverage. The problems will be familiar to the cellular operators, who expend considerable time and effort planning their networks to use the minimum number of base stations for the required coverage. Cellular operators invariably deploy planning tools to aid them in this task.

Planning tools basically are computer programs that use as an input a digital map of the area. Using algorithms derived from propagation rules, coverage areas for hypothetical base station sites can be predicted. Different base station locations can be tried rapidly until a near-optimal result is achieved. Planning tools are available for cellular planning from a wide variety of vendors covering different technologies and cell sizes.

WLL operators need to use planning tools in the same manner as cellular operators. However, as discussed in this chapter, the propagation modes for WLL are significantly different from cellular, particularly due to the high reliance on a LOS path in WLL systems. No cellular models provide adequate prediction of LOS paths, considering shadowing from high buildings as well as terrain. Therefore, new tools are required to plan WLL systems. At the time of this writing, no such tools are available on the market, but some organizations are in the last stages of development of appropriate tools that are expected to be introduced during 1997.

When such tools are available, the WLL operator need not overly concern themselves with how they work. For those who are interested, the general principles are as follows:

- The tool starts with a digital map of the target area. The map needs to show the outlines of every building and the building heights. Such maps are available from aerial photography and increasingly from satellite photography. Depending on the coverage area required, maps can be expected to cost around $200 to $300 per square km.

- The user then selects a particular building or mast on which to place a hypothetical base station. The program has an algorithm that completes a 360 degree scan from the selected base station site. For each point on the scan, the program needs to consider azimuth angles from, say, +20 to −70 degrees. For each azimuth point, a "ray" is sent from the base station. The buildings that the ray strikes have a LOS path, and the path loss can be calculated based simply on the distance and the path-loss law used by the model. When a ray grazes the top or side edge of a building, diffracted rays are generated, which are then followed until they strike further buildings. The path loss on those rays also needs to include the diffraction loss.

■ A further enhancement also can allow for reflected signals by following the path of a ray as it is reflected off a building. This enhancement provides a slight improvement in accuracy but at the expense of a substantial increase in computing requirements. Such an enhancement typically is not necessary.

A WLL operator will need to select and purchase an appropriate planning tool prior to deploying a network. Up-to-date information about planning tools can be obtained from information sources in trade journals or from those working in the industry. Until such tools are released, no further guidance can be provided here.

Chapter 7 looks at the design of a generic radio system, which will be of use in understanding the advantages of particular radio systems. It also will show how radio designers overcome some of the propagation problems described in this chapter.

References

[1]   Steele, R., ed., *Mobile Radio Communications*, New York: Wiley & Sons, 1992.

[2]   Freeman, R., *Radio System Design for Telecommunications (1–100 GHz)*, New York: Wiley & Sons, 1987.

[3]   Withers, *Radio Spectrum Management*, London: IEE Press, 1987.

7

Radio Systems

THIS CHAPTER EXPLAINS the general design philosophies of radio systems, concentrating on those parts of the radio system that are of relevance to WLL technology. Figure 7.1 is a block diagram of a stylized digital radio system. Note that the ordering of blocks is important: speech coding must be performed first, modulation and medium access last, and ciphering before error correction. Analog radio systems, broadly speaking, do not have any of those steps and are sufficiently simple not to need further discussion here.

Each of the stages is now described in more detail.

7.1 Speech coding

In digital radio systems, it is necessary to turn voice signals, which are analog, into a digital data stream. (If data rather than voice is to be transmitted, this stage is not required.) Speech coding is a highly complex

89

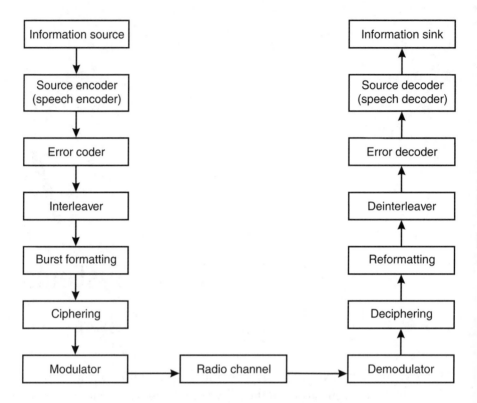

Figure 7.1 Block diagram of a radio system.

topic; thousands of researchers around the world are working on it, and the area is evolving daily. Readers who want to know more need to consult relevant journals; a suitable starting point can be found in [1]. Here, a simplified treatment is provided for two reasons: (1) a full treatment is beyond the scope of this book, and (2) as will become evident, the more advanced speech coders currently are not appropriate for WLL networks.

The simplest speech coders essentially are analog-to-digital converters. The analog speech waveform is sampled periodically, and the instantaneous voltage level associated with the speech converted to a digital level. Figure 7.2 is a stylized diagram that illustrates the process.

The amount of information transmitted and the quality of the speech depend on two parameters. The first parameter is how frequently the speech is to be sampled, that is, the "sampling rate." The second

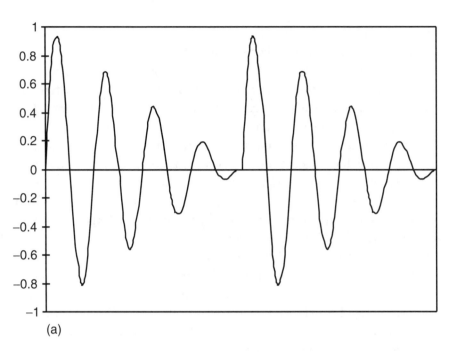

(a)

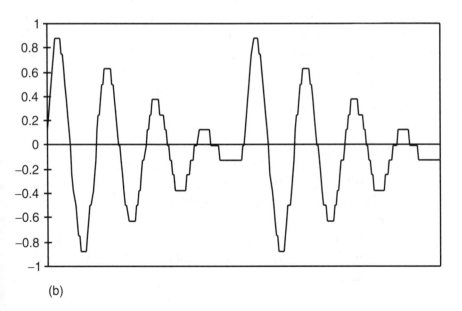

(b)

Figure 7.2 Speech waveforms (a) before and (b) after A-D conversion.

parameter is how many bits are used to describe the voltage level: the more the bits, the less the resulting waveform will look "step-like." In technical terms, the difference between the digital waveform and the original analog waveform is known as the quantization error. The tradeof is to provide a sufficiently high sampling rate and number of quantization bits to provide good voice quality while avoiding using so many tha spectrum resources are wasted. In calculating the sampling rate, a theo rem known as the Nyquist theorem can be used. The Nyquist theorem states that if a waveform is sampled at twice the highest frequency in the waveform it can be correctly re-created. Voice generally is assumed no to contain any useful information above frequencies of 4 kHz[1]. Hence, a sampling rate of 8 kHz typically is sufficient for an acceptable voice quality (at least equivalent to existing telephone systems).

There is an internationally agreed standard for voice coding using 8-kHz sampling, known as *pulse code modulation* (PCM). PCM utilizes 8-bit sampling at 8 kHz, resulting in a bit rate of 64 Kbps. This voice coding system currently is used in most network backhaul links and digita switches. Its voice quality is considered to be near perfect.

From an information viewpoint, PCM is extremely inefficient because it takes no account of the periodicity and predictability in speech. Typically, over the period of one syllable, lasting perhaps 20 ms, the speech is unchanging and can be represented better by a periodicity and a waveform. More generally, speech waveforms are smooth: changes to different syllables occur slowly. Those facts can be used to reduce the bandwidth of the digital signal.

The simplest system that starts to use the fact that much of the voice waveform is predictable is known as *adaptive differential PCM* (ADPCM). Like PCM, ADPCM samples the speech waveform at 8 kHz. However, instead of quantifying the absolute signal level, ADPCM quantizes the difference between the signal level at the previous sample and the current sample. Because speech changes slowly, that difference typically is small and hence can be accurately represented by fewer bits than the absolute level. That is the differential part of ADPCM. A further

---

1. That is not strictly true. Some sounds like "s" contain substantial energy above 4 kHz. The removal of such frequencies in telephone systems can make some sounds difficult to understand, particularly isolated letters.

enhancement is that when speech waveforms are slowly changing it tends to occur for some time, so during that time the size of the gap between each of the quantization levels can be reduced. That gap is increased as the waveform changes more quickly. Hence, the quantization size becomes adaptive to the waveform.

ADPCM generally works well. However, because the differential is transmitted, an error in receiving the differential makes the receiver represent a different point on the received waveform than that transmitted. That difference continues while only differences are sent. Further, an error in reception could lead to the receiver changing the adaptive levels to different settings than the transmitter, resulting in further errors. Those errors propagate until the transmitter sends a reset-type signal, resynchronizing the receiver. That needs to be performed regularly to prevent error propagation from becoming severe. Even with regular transmission, the effect of an error can be a short glitch in the speech. These types of error are the price paid for reducing the information used to transmit the signal.

Because ADPCM systems use only 4 bits to quantize the difference in the waveform, the bitstream generated is 32 Kbps, only half that of PCM. ADPCM generally is considered to provide quality virtually as good as PCM and is widely used for WLL systems.

Information theory shows that it is possible to go much further. More optimal voice coders would model the vocal tract of the speaker based on the first few syllables and then send information on how the vocal tract is being excited. The speech coders used by digital cellular systems follow that route to a greater or lesser extent. For example, GSM utilizes a coder type known as *regular pulse excited–long term prediction* (RPE-LTP), where the LTP part sends some parameters showing what the vocal tract is doing, and the RPE shows how it is being excited. Such speech coders are extremely complex and beyond the scope of this book.

As far as the WLL operator is concerned, the key parameters are the bandwidth required, which has direct implications on the number of subscribers that can be supported, and the voice quality. Table 7.1 lists six voice coders and general perceptions about their quality.

For the WLL operator, voice quality can be particularly important. While users accept, unwillingly, that voice quality in a mobile radio system can be inferior to wireline quality, that typically is not the case for

Table 7.1
Six Voice Coders

Codec	Data Rate	Quality	Comments
PCM	64 Kbps	Excellent	
ADPCM	32 Kbps	Virtually PCM	WLL applications
GSM full rate	13 Kbps	Noticeably worse than PCM	
Enhanced GSM full rate	13 Kbps	Better than GSM full rate	To be introduced soon
GSM half rate	6.5 Kbps	Slightly worse than GSM full rate	Unlikely to be widely used
TETRA	4.4 Kbps	To be demonstrated	

a WLL operator. Further, the WLL operator may be less constrained by spectrum than the mobile operator. The net result is that WLL operators tend to use coders with better voice quality than those used by cellular operators, by far the majority opting for the ADPCM coder.

It is certain that coders will improve relatively quickly in the coming years. A GSM-enhanced full-rate coder, operating at 13 Kbps but with enhanced speech quality, is to be introduced soon, and other improvements can be expected. It is likely that the data rate required for WLL speech will fall to perhaps 16 Kbps in three years and probably as low as 8 Kbps in 5 to 10 years. Further advances beyond that would be entirely possible.

7.2 Ciphering

Almost all modern radio systems rely on some form of ciphering to provide secure transmission. The use of such security techniques both reassures the user that the conversation cannot be overheard and allows the operator to authenticate the user.

Most ciphering schemes are relatively straightforward. The base station and the subscriber unit agree on a "mask" on a call-by-call basis.

Such a mask might be 32 or 64 bits long. At the transmitter, the data to be sent is multiplied by the mask (strictly speaking, the operation used is a binary XOR operation), as shown in Figure 7.3.

The receiver can then multiply (XOR) the received data by the same mask to decode the transmitted data. To a covert listener, who does not know the mask, the data is incomprehensible.

The difficulty is in the process by which the transmitter and the receiver agree on the mask. The mask simply cannot be transmitted at the start of the call because a covert listener could receive the mask and apply it in the same way as the receiver. Instead, the mask must be generated using information available only to the transmitter and the receiver.

At manufacture, the receiver is given a unit code number. The number is kept highly confidential and is known only to the network. It is never transmitted over the air. At the start of a call, the network sends the receiver a random number. The receiver combines the random number with the secret unit code using a special algorithm, the output of which is the mask. A covert listener does not know the unit code and so is unable to generate the mask. The use of a different random number from call to call makes it more difficult for a covert listener to break the mask using statistical processes.

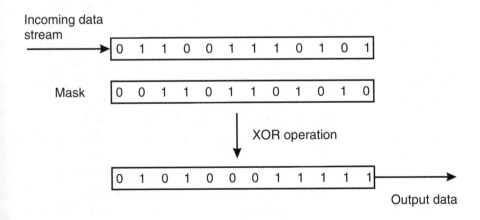

Figure 7.3 Multiplication by a mask.

7.3 Error-correction coding and interleaving

Information transmitted via a radio channel is liable to be corrupted. Interference, fading, and random noise cause errors to be received, the level of which depend on the severity of the interference. The presence of errors can cause problems. For speech coders such as ADPCM, if the BER rises above 10^{-3} (i.e., 1 bit in every 1,000 is in error, or the error rate is 0.1%), the speech quality becomes unacceptable. For near-perfect voice quality, error rates of the order of 10^{-6} are required. For data transfers, users expect much better error rates, for example, on computer files, and error rates higher than 10^{-9} normally are unacceptable.

If the only source of error on the channel was random noise, it would be possible—and generally efficient—simply to ensure that the received signal power was sufficient to achieve the required error performance without any need for error correction. However, where fading is present, fades can be momentarily as deep as 40 dB. Increasing the received power by 40 dB to overcome such fades would be highly inefficient, resulting in significantly reduced range and increased interference to other cells. Instead, error-correction coding accepts that bits will be received in error during fades but attempts to correct those errors using redundancy added to the signal.

Error correction is widely deployed in mobile radio, where fast fading is almost universally present. It is less critical in WLL, where the LOS path and resultant amelioration in the fast fading cause less severe problems. Nevertheless, it is still necessary, especially for computer file transfer. Further, because many of the WLL technologies are based on cellular standards, they tend to come with the error correction used for cellular already inbuilt. Some WLL systems go so far as to use half-rate error-correction coding, that is, twice as much information as required is transmitted to introduce the required redundancy for the powerful error correction specified.

Error-correction systems work by adding redundancy to the transmitted signal. The receiver checks that the redundant information is as it would have expected; if it is not, the receiver can make error-correction decisions. An extremely simple error-correction scheme repeats the data three times. The first bit in each of the three repetitions is compared; if there is any difference, the value that is present in two of the three repetitions is assumed to be correct. The procedure is repeated for all bits. Such a system corrects one error in every three bits but triples the bandwidth required. Considerably more efficient schemes are available.

There are schemes that detect errors but do not correct them. In the simple example in the preceding paragraph, if the message is repeated only twice and the repetition of a given bit is not the same as the original transmission, it is clear that an error has occurred, but it is not possible to say which transmission is in error. In an error-detection scheme, the receiver requests that the block that was detected to be in error is retransmitted. Such a scheme is called an *automatic repeat request* (ARQ). ARQs have the advantage of often reducing the transmission require-ments (even accounting for the bandwidth needed for retransmission of errored blocks) but add a variable delay to the transmission while blocks are repeated. The variable delay is unsuitable for speech but typically acceptable on computer file transfer. Some of the more advanced coding systems can perform error correction and also detect if there are too many errors for it to be possible to correct them all and hence request retransmission.

Error-correction methodologies broadly fall into two types, block coding or convolutional coding. Both are highly involved and mathemati-cal, and the treatment here will no more than scratch the surface. Interested readers are referred to [1]. Block coding works basically by putting the information to be transmitted in a matrix and multiplying that by another matrix, whose contents are fixed for the particular coding scheme and known to both the transmitter and the receiver. The result of the matrix multiplication forms the codeword. The codeword then is transmitted after the information, which is left unchanged. At the re-ceiver, the information is loaded into another identical matrix and mul-tiplied by the known matrix, and the results are compared with the received codeword. If there are differences, complex matrix operations (which are processor intensive) can be used to determine where the error lies and it can be corrected. If no solution can be found, more errors than can be corrected have occurred.

Depending on the type of matrix, codes fall into a number of families. Two well-known families are the *Bose-Chaudhuri-Hocquenghem* (BCH) family and the *Reed-Solomon* (RS) family. The coding power is represented by a shorthand, for example, BCH(63,45,3). That code takes a block of 45 bits, adds 28 coding bits, and then transmits the 63-bit signal. It can correct up to 3 errors in the received signal (an error rate of $3/63 = 4.7\%$), but it increases the datastream to be transmitted by $63/45 = 1.4$, or 40%.

Convolutional coders are completely different. Figure 7.4 is a diagram of a highly simplified convolutional coder.

By combining the input and previously input bits in a variety of ways, predictable redundancy is added to the signal. For example, in Figure 7.4, where the plus sign represents an XOR operation, if the previous three bits had been 010 and then a new bit arrived, the three bits in the encoder would be either 101 or 001, depending on whether the new bit was a 1 or a 0, respectively, and the only possible outputs would be 01 for bits 1 and 2, respectively, or 11 (if 00 or 10 was received, there would have been an error). The decoder in the receiver can use that knowledge of the redundancy to correct errors. It does that by using an architecture known as a Viterbi decoder.[2] Such decoders are used widely in mobile radio, but they are complex to describe and beyond the scope of this text.

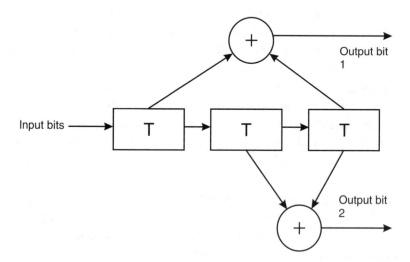

Figure 7.4 Diagram of a convolutional encoder.

---

2. The Viterbi decoder is named after Andrew Viterbi, one of the cofounders of Qualcomm, a WLL equipment manufacturer.

The references provide a full description. Put briefly, a Viterbi decoder computes what it would have received for all possible inputs (in the case of the example, a total of $8 = 2^3$ different inputs). It then compares the received signals with each of those possible inputs and scores each possible input depending on how close the input is to the received signal. When the next symbol arrives, it does the same thing. Also, for each of the inputs for the first bit the decoder calculates which possible inputs could be expected if either a 0 or a 1 was inserted as the next bit (as explained at the start of this paragraph). It then adds to the score of the first input the score of each of the two possible next symbols, building a path through the received data.

Because the original data is never transmitted, unlike block codes, convolutional codes cannot perform error detection. Therefore, they cannot deploy ARQ. However, compared to block codes, they both are more powerful and require less computing power. Often radio systems use both types of error-correction systems in a concatenated arrangement to get the best of both worlds.

All error-correction systems work best when the errors are randomly distributed and worst when the errors arrive in blocks. That can be readily seen with the BCH(63,45,3) example. Ideally, in each block of 63 bits, there would be three (or fewer errors). The system will not work if there are no errors in the first 3 blocks and then 12 errors in the next block. Unfortunately, fading tends to result in errors occurring in blocks. The solution is to randomize the errors using a device known as an interleaver. Interleavers are quite simple devices. A typical interleaver would place the input bits in a matrix, filling it from left to right and then moving down a row. When the matrix is full, the data is read out in columns and transmitted. At the receiver, the data fills up a matrix from top to bottom in columns and when full reads it out in rows from left to right, as shown diagrammatically in Figure 7.5.

A block of errors would affect one column, but only one error would appear on each row, thus reducing the impact of fades.

Interleavers have a key disadvantage in that they introduce delay while the matrix at the transmitter is filled and while the matrix at the receiver is filled. That delay is undesirable for speech. System designers need to balance the reduction in error rate (and hence improvement in speech quality) with the undesirable effect of the delay.

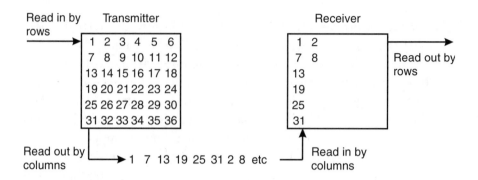

Figure 7.5   Interleaving.

7.4 Modulation

In a radio system, there is an encoded and error-corrected digital data stream consisting of square-wave binary pulses. If such a waveform were transmitted directly, it would use vast amounts of spectrum. Fourier transform theory shows that an instantaneous change in the time domain (e.g., the rising edge of a binary pulse) requires an infinite amount of bandwidth in the frequency domain. Something more needs to be performed before the signal can be transmitted, and that something is modulation.

There are two different classes of modulation, *phase modulation* (PM) and *amplitude modulation* (AM). Frequency modulation is strictly a class of PM and is not discussed further. Broadly defined, PM changes the phase of the transmitted signal in accordance with the information to be transmitted, and AM changes the amplitude of the transmitted signal. Example waveforms for the two types of modulation are shown in Figure 7.6. Typically, PM is less susceptible to interference and is more widely used.

It is possible to combine AM and PM; the result is QAM (*quadrature amplitude modulation*). As yet, no mobile radio system or WLL system has been developed with QAM, so it is not discussed in more detail. It is worth noting that QAM is used widely on twisted pair and cable along with complex derivatives such as *trellis code modulation* (TCM). Those interested in finding out more about QAM should refer to [2].

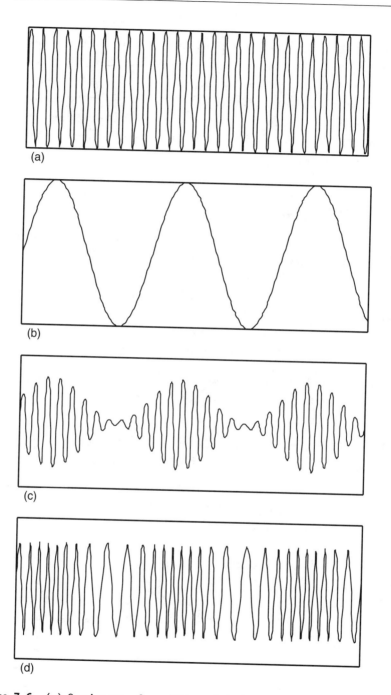

Figure 7.6 (a) Carrier waveform; (b) modulating waveform; (c) amplitude-modulated waveform; (d) frequency-modulated waveform.

Before the square baseband pulses are used to modulate the carrier, the sharp changes must be removed. That is achieved with the use of pulse shaping. The square pulses are passed through a filter, resulting in rounded pulses. The more severe the filter, the lower the bandwidth of the transmitted signal, but the more the difficulty in decoding the signal. Depending on the bandwidth scarcity, such tradeoffs need to be made by the system designer.

7.5 Multiple access

For most of the topics in Part III, information is provided mostly for background knowledge. Multiple access, however, is of greater importance because the available WLL technologies differ in the multiple-access method they use. Decisions about which technology to adopt will be influenced significantly by the access medium required.

Each operator has a given amount of radio spectrum to divide among its users. There are broadly three ways to do that:

■ *Frequency division multiple access* (FDMA), in which the frequency is divided into a number of slots and each user accesses a particular slot for the length of the call;

■ *Time division multiple access* (TDMA), in which each user accesses all the frequency but for only a short period of time;

■ *Code division multiple access* (CDMA), in which each user accesses all the frequency for all the time but distinguishes the transmission through the use of a particular code.

Figure 7.7 shows the different multiple-access methods in a diagrammatic form.

WLL technologies are available that make use of each of the different access methods. Each method and its advantages and disadvantages are described next in more detail.

7.5.1 FDMA

In a typical FDMA system, the available bandwidth is divided into slots about 25 kHz wide as shown diagrammatically in Figure 7.8. Each slot

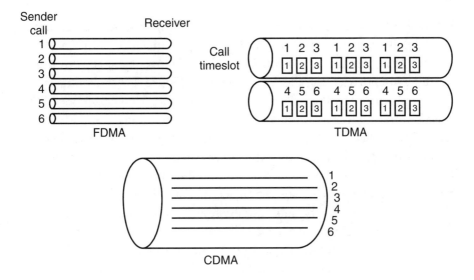

Figure 7.7 Diagrams of FDMA, TDMA, and CDMA access methods.

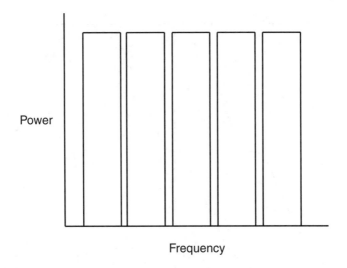

Figure 7.8 An ideal FDMA spectrum.

contains a dedicated transmission. The problem lies in the fact that the transmitted power when plotted against bandwidth is not an idealized rectangle (i.e., equal power is not transmitted across the entire available bandwidth). To generate such a transmission would require that the input signal be filtered over an infinitely long time period, clearly not a practical option. With the use of appropriate pulse shaping, the spectrum transmitted can approach the ideal requirement. Figure 7.9 shows the spectrum transmitted by the GSM mobile radio system, which uses one of the most complex filtering arrangements of any technology currently available. Two adjacent channels are shown, and it is clear that there is significant interference between the two channels. Because of that, it is not possible to use adjacent channels in the same cell. Had the channels been spaced farther apart so that adjacent channels could be used in the same cell, there would be substantially fewer carriers per megahertz and the spectrum efficiency of the overall system would be much lower.

It is clear that substantial energy extends outside the channel in which a single user is transmitting. The result is that adjacent channels cannot be used in the same cell because the interference between them would be too great. It is also clear that, compared to an ideal rectangular spectrum emission, the GSM system does not transmit as much power

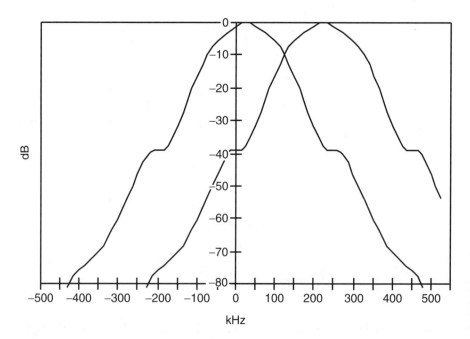

Figure 7.9 GSM spectrum.

within the band; the power levels fall away as the band edges are reached. That represents a lost opportunity, reducing from the ideal case the power available to the mobile.

A more pragmatic problem with FDMA is that at the base station each individual channel requires a separate power amplifier before passing through an expensive high-power combiner and then being transmitted from the antenna. It is possible to combine the signals before the amplifier if the amplifier is highly linear, but in practice such amplifiers are extremely expensive and inefficient in their use of power.

In summary, the advantage of FDMA is that it is the simplest access method to implement. The disadvantages are the following:

- The loss of efficiency caused by imperfect filtering;

- The expensive *radio frequency* (RF) elements required at the base station.

Now that technology has advanced to where other access methods can be implemented at relatively low cost, the single advantage of FDMA is broadly negated. For that reason, virtually no digital WLL systems use FDMA, and it is in use only in the older analog WLL technologies.

7.5.2 TDMA

In TDMA, a user has access to a wide bandwidth but only for a short period of time. Using the example of GSM, which is a TDMA system, a user has access to 200 kHz of bandwidth for one-eighth of the time. To be more precise, the user has access to the channel for 577 μs every 4.6 ms. During that period, the transmitter sends a burst of data that was previously buffered in the transmitter.

Like FDMA, TDMA has its inefficiencies. Those inefficiencies are caused by the need to allow the mobile time to increase its power from zero and to reduce it back to zero again. If time is not allowed for that transition, the near-instantaneous change in power is, in effect, close to transmitting a square wave, resulting in a momentary use of an extremely high bandwidth, with resulting interference to a wide range of users. Guardbands are provided to allow for the powering up and down. The structure of a burst in TDMA is shown in Figure 7.10, where it can be

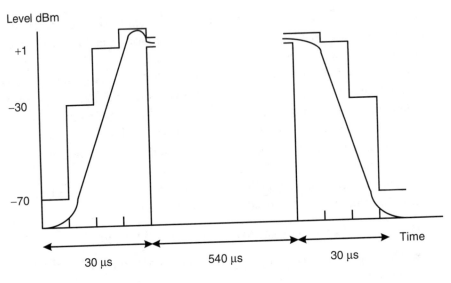

Figure 7.10 GSM ramping up for a burst.

seen that around 30 μs is required to ramp up for a burst of 540 μs. (The time taken to ramp down also is used by the next mobile to ramp up, so it is not required in addition to the ramp-up time.) Hence, the inefficiency is around $30/540 = 5.5\%$.

TDMA systems require additional overhead because they have to send timing information so the subscriber units know exactly when to transmit. There is an additional problem. By transmitting over a wider bandwidth, the problem of ISI is exacerbated. In some systems, such as GSM, that has resulted in the need for a component called an equalizer, which removes the ISI in the receiver. For an equalizer to work, the channel impulse response needs to be measured periodically. The timing is performed by placement of a training sequence in the middle of each burst. By correlating the received training sequence with a local copy of the training sequence, the channel impulse response can be measured. However, the sending of the training sequence represents a substantial inefficiency in the use of radio spectrum. Training sequences can be avoided for WLL because, as explained in Chapter 6, the ISI is reduced through the use of directional antennas at the subscriber units. Nevertheless, care must be taken because it cannot be guaranteed that ISI will not be problematic in wideband systems.

TDMA overcomes one key disadvantage of FDMA. Because only one user's signal is transmitted at any one time, only a single amplifier is required at the base station and the need for combiners is removed. That provides a substantial cost saving in the base station. TDMA also is, generally, more spectrum efficient than FDMA, because the size of the guardband relative to the size of a burst is much smaller than the size of the guardband effectively required for FDMA relative to the bandwidth of an FDMA channel.

In summary, TDMA has the following advantages over FDMA:

- It is more efficient.

- It is less expensive to implement.

The disadvantages of TDMA are the following:

- More complex subscriber units are required.

- ISI may become problematic.

TDMA is a widely used multiple-access method for many mobile and WLL systems. Strictly speaking, most TDMA systems actually are TDMA/FDMA. For example, GSM places eight users on a 200-kHz channel using TDMA but then divides the assignment into 200-kHz slots using FDMA.

7.5.3 CDMA

This section goes into somewhat more detail, because CDMA is more complex and less intuitive than the other access technologies. There is significant debate as to whether CDMA or TDMA is more appropriate, and an understanding of that debate requires a good understanding of CDMA itself.

CDMA is the process of data transmission using a code. In that process, each user is allocated a particular codeword. The user first generates data, which could be, for example, the output of a speech coder. The data is generated at a rate known as the bit rate, or R_b. Each bit is multiplied by the code to achieve the final output stream. The output stream is at a data rate equal to R_b multiplied by the length of the

codeword, G; that rate is the chip rate, or R_c. The process of multiplication by a codeword is known as spreading and is shown for an example
datastream and codeword in Figure 7.11.

The length of the codeword, and hence the chip rate, is a fundamental
design parameter of a CDMA system. Spreading with a large codeword
results in a large transmitted bandwidth, which, if there are only a small
number of users, will prove inefficient. Spreading with a smaller codeword yields a smaller transmitted bandwidth, which may not be able to
accommodate sufficient users. The spreading factor also is influenced by
the size of the frequency assignment available to the operator.

The choice of the code is critically important. The code should have
good autocorrelation properties such that when correlated with offset
versions of itself it exhibits a large impulselike peak at zero offset and a
small residual signal at other offsets. Such a property maximizes the
probability of reception. Many families of such sequences are known.

At the receiver a process of despreading is required to recover the
data. The process involves the multiplication of the received signal with
the codeword. Such multiplication results in the original binary information being decoded but with an enhancement of the signal level by a factor
of G. The enhancement allows interference to be tolerated on the link.
For example, if the speech coder required an SNR of 9 dB (a factor of 8)
and the code length G was 64 bits, then a signal of $8/64 = 0.12 = -9$ dB
could be tolerated. In that instance, the wanted signal is at a lower level
than the interference, but it can be decoded because of the wider
bandwidth it occupies.

In CDMA WLL systems, the ability to tolerate interference is used
to allow other users to send their transmissions on the same channel. Each
of the other users also has a spreading code. It is important that each user
have a different code and that the codes are orthogonal, or nearly so, with
each other. If two users had the same code, the receiver would not be
able to differentiate between them and the interference would be severe.
If two users employed codes that were different but not orthogonal, a
certain component of the second user's signal would be decoded by the
first user, negating some of the advantages of the decoding process. A
number of code families are orthogonal; the best known probably is the
Walsh family. In any one family, there can be only as many orthogonal
codewords as there are bits in the codeword, that is, G.

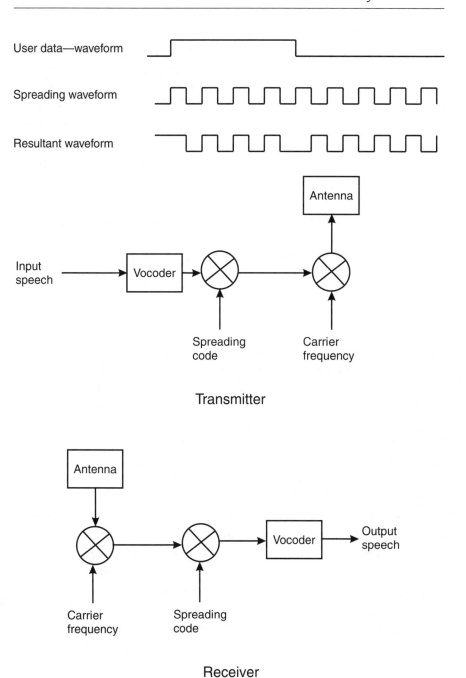

Figure 7.11 Example of the generation of a CDMA signal.

The shortage of codewords can be a problem in some systems particularly cellular, where interference from adjacent cells can be expected. It is overcome by using nearly orthogonal code families. In such families, there is no shortage of codes, but the interference from other users becomes slightly more problematic. It is worth going into this topic in a little more detail because the allocation of codewords has become a parameter that an operator needs to define in some CDMA WLL systems. Codewords are explained further in Section 7.6.

Based on the discussion so far, it is apparent that the capacity of a CDMA system is limited by the amount of interference generated by other users employing the same frequency. As such, it often is stated that "CDMA is interference limited." While in a WLL system, all access schemes ultimately are interference limited, the link between CDMA capacity and interference is more direct and apparent than in other access techniques.

One of the advantages of CDMA is that by reducing the spreading factor G, a user can increase the transmitted data rate without changing the bandwidth of the signal being transmitted. That allows a form of dynamic bandwidth-on-demand allocation. However, for the user who reduces the spreading factor, the interference that can be tolerated reduces, and the overall system capacity in terms of number of users falls. Nevertheless, such flexibility is extremely useful in the WLL environment.

One of the main concerns with CDMA is power control. In the example given above, where G = 64 and an SNR of −9 dB could be tolerated, if the signal from the wanted mobile is at a level S, the total received signal power, including interference, would be 8S. Therefore seven other users transmitting the same signal level S could be supported. If, however, one user transmits at a level 4S, then only four other users could be supported. In a WLL environment, some subscriber terminals are closer to the base station than others. To maximize the system capacity, it is important that those terminals closer to the base station transmit with a lower power so that all signal levels are received with the same signal strength.

The accuracy required for power control is high. A 3-dB error in single subscriber unit would halve the capacity of the system. In cellular

CDMA systems, such accuracy is difficult to achieve as mobiles pass through a fast-fading environment. With WLL, the situation is less problematic, as was explained in Chapter 6. In particular, where there is a LOS path, the channel will be relatively stable, and closed-loop power control systems can be used to accurately trim the power level to exactly that required. The effects of imperfections in the average power control accuracy for all the subscriber units are shown in Figure 7.12.

Within the mobile and WLL community, there has been much impassioned discussion about whether TDMA or CDMA represents the best access technique. Because of the importance of this debate, the whole of Chapter 8 has been set aside to address it.

7.6 The use of spreading codes in a CDMA system

As a simple introduction to CDMA systems, Section 7.5 simplified the design issues associated with spreading codes. This topic is important for operators planning to deploy WLL systems based on CDMA because they will need to assign families of spreading codes to particular cells.

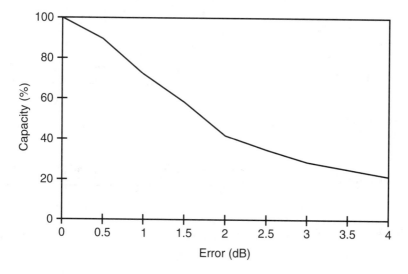

Figure 7.12 Sensitivity to power-control error in CDMA systems.

In an isolated CDMA cell, each user's signal is spread by a particular spreading code. Spreading codes should be orthogonal so they do not generate interference. Two codes are orthogonal to each other if, when they are multiplied over the length of the spreading code and the total summed, the result is zero, regardless of whether the code was carrying user data corresponding to a 1 or a 0. That is best illustrated by a sequence of examples. To allow the sequence to fit on the page, a spreading factor of 8 has been used, that is, for every bit of user information, 8 bits of the spreading code are transmitted. In practice, a spreading factor of typically around 64 would be used. In the examples, the spreading waveforms have been shown as sequences of 1s and −1s, while the user data has been shown as 0s and 1s. This is shown diagrammatically in Figure 7.13.

Moving to this numerical representation, it is now possible to examine how orthogonal codes work in more detail. First, consider how the receiver decodes the wanted signal by multiplying the received signal by a copy of the spreading sequence held in the receiver. Assume that the spreading sequence is 1, −1, 1, −1, 1, −1, 1, −1, as shown in Figure 7.13.

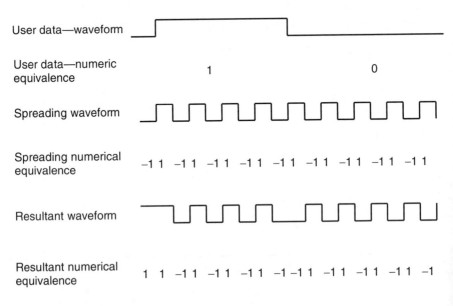

Figure 7.13   Numeric representation of CDMA waveforms.

Transmit (data = 1)	1	−1	1	−1	1	−1	1	−1	
Receiver codeword	1	−1	1	−1	1	−1	1	−1	
Multiplication	1	1	1	1	1	1	1	1	= 8

Transmit (data = 0)	−1	1	−1	1	−1	1	−1	1	
Receiver codeword	1	−1	1	−1	1	−1	1	−1	
Multiplication	−1	−1	−1	−1	−1	−1	−1	−1	= −8

The effect is that when the received waveform is multiplied by the spreading code and integrated over the period of one bit of user data, the output of the correlator is either 8 or −8, corresponding to a 1 or a 0 in the user datastream. (If a spreading code of length 64 had been used, the output would have been either 64 or −64.)

Now it so happens that a spreading sequence of 1, 1, −1, −1, 1, 1, −1, −1 (i.e., a square wave with half the periodicity of our spreading sequence) is orthogonal to our spreading sequence. Using the same multiplication process, but now assuming that two subscribers are transmitting at the same time, the following multiplication occurs in the base station.

Transmit (data = 1)	1	−1	1	−1	1	−1	1	−1	
Interferer (data = 1)	1	1	−1	−1	1	1	−1	−1	
Received signal	2	0	0	−2	2	0	0	−2	
Receiver codeword	1	−1	1	−1	1	−1	1	−1	
Multiplication	2	0	0	2	2	0	0	2	= 8

Transmit (data = 0)	−1	1	−1	1	−1	1	−1	1	
Interferer (data = 1)	1	1	−1	−1	1	1	−1	−1	
Received signal	0	2	−2	0	0	2	−2	0	
Receiver codeword	1	−1	1	−1	1	−1	1	−1	
Multiplication	0	−2	−2	0	0	−2	−2	0	= −8

It can be seen that the output from the correlator is the same despite the fact that an interfering signal was on the channel. Next, consider the situation in which the interferer is sending user data corresponding to a 0.

Transmit (data = 1)	1	−1	1	−1	1	−1	1	−1	
Interferer (data = 0)	−1	−1	1	1	−1	−1	1	1	
Received signal	0	−2	2	0	0	−2	2	0	
Receiver codeword	1	−1	1	−1	1	−1	1	−1	
Multiplication	0	2	2	0	0	2	2	0	= 8

Transmit (data = 0)	−1	1	−1	1	−1	1	−1	1	
Interferer (data = 0)	−1	−1	1	1	−1	−1	1	1	
Received signal	−2	0	0	2	−2	0	0	2	
Receiver codeword	1	−1	1	−1	1	−1	1	−1	
Multiplication	−2	0	0	−2	−2	0	0	−2	= −8

So whatever is transmitted by the interferer and by the wanted user, the correlator produces the same result as if there were no interferer. Clearly the signals are orthogonal. More orthogonal interferers can be added without having any effect on the wanted signal. Even if the interferers are received with different power from the transmitter, the correct result still is achieved, as demonstrated next.

Transmit (data = 1)	1	−1	1	−1	1	−1	1	−1	
Interferer (data = 1)	2	2	−2	−2	2	2	−2	−2	
Received signal	3	1	−1	−3	3	1	−1	−3	
Receiver codeword	1	−1	1	−1	1	−1	1	−1	
Multiplication	3	−1	−1	3	3	−1	−1	3	= 8

Transmit (data = 0)	−1	1	−1	1	−1	1	−1	1	
Interferer (data = 1)	2	2	−2	−2	2	2	−2	−2	
Received signal	1	3	−3	−1	1	3	−3	−1	
Receiver codeword	1	−1	1	−1	1	−1	1	−1	
Multiplication	1	−3	−3	1	1	−3	−3	1	= −8

The only time this relationship does not hold is if the received signals are not synchronized so that the transitions in the spreading sequences occur at different times. That would occur if the subscriber units were different distances from the base station and so experiencing different propagation delays. That situation can be ameliorated through the use of timing advance commands from the base station to tell the subscriber unit to change its internal clock by the propagation delay being experienced.

This is one of the key differences between the use of CDMA for mobile and WLL purposes. In the mobile case, synchronization on the uplink is almost impossible to achieve because of the movement of the mobile, resulting in variable propagation delays. Hence, the uplink is designed to accept asynchronous input, resulting in a lower system capacity. In WLL systems, there is no movement; hence, through the use of timing advance mechanisms when the subscriber unit is set up, a synchronous system can be produced. That allows a lesser reliance on *pseudo-noise* (PN) codes (which are more tolerant to asynchronization than Walsh codes), with resulting lower interference and greater capacity. CDMA actually is better suited for WLL applications than it is for mobile applications.

These codes are actually two of the so-called Walsh code family. The complete family is given by:

Code 0	1	1	1	1	1	1	1	1
Code 1	1	1	1	1	−1	−1	−1	−1
Code 2	1	1	−1	−1	−1	−1	1	1
Code 3	−1	−1	1	1	−1	−1	1	1
Code 4	1	−1	−1	1	1	−1	−1	1
Code 5	1	−1	−1	1	−1	1	1	−1
Code 6	1	−1	1	−1	1	−1	1	−1
Code 7	−1	1	−1	1	1	−1	1	−1

All the codes are orthogonal to each other, which can be demonstrated by multiplying any code by any other code; the result will always be 0.

In a single cell in isolation, each channel can be given a separate Walsh code and the maximum system capacity can be reached. The problem comes when a number of neighboring cells use the same frequency. To prevent excessive interference, all the users in all the neighboring cells need different spreading sequences. However, there are only as many sequences as the code length, whereas there will be many more interfering users than that. Of course, one solution would be not to let neighboring cells use the same frequency, but the result of that would be a much lower system capacity than could otherwise be the case.

For a long time, researchers thought that the problem meant that CDMA could not be used for commercial applications. The breakthrough came with the realization that if nearly orthogonal codes were used rather than fully orthogonal codes the interference between different users would not be too severe, although greater than for the orthogonal case. A family of near-orthogonal codes is PN codes. The good thing about PN codes is that there literally are millions of them, overcoming the limitations of Walsh codes.

A PN code is a sequence of 1s and 0s that repeats periodically. PN code sequences have the property that if multiplied by themselves the result has the same magnitude as the length of the sequence (in the same way that the orthogonal codes resulted in +8 or −8). If multiplied by the same sequence but shifted in time by any number of bits, the result is −1 (unlike orthogonal codes, where the result was 0).

A number of different approaches have been taken to the use of Walsh and PN codes in commercial mobile CDMA systems and in WLL systems. In the mobile systems, there is a tendency to give users separate PN codes, since they could roam into any cell and there is a need to ensure that they do not generate interference wherever they are. In WLL systems, a simpler approach can be adopted, since a subscriber unit always uses a particular cell.

The approach adopted for WLL systems in one particular cell has been to use the Walsh sequences to spread the signals but then to multiply the signal by a PN code at the same rate as the Walsh code. That does not result in any further spreading, because the data rates are the same. Indeed, multiplying all the transmitted sequences by the same PN code does not affect their orthogonality (although multiplying each by a different PN code would). Hence, within a particular cell, the performance is unchanged. In a neighboring cell, the same approach is adopted but with a different PN code. Thus, users in neighboring cells might be using the same Walsh code but different PN codes. The result is that the users do not interfere severely, as would be the case if PN codes were not used. However, because the PN codes are not orthogonal, there is limited interference between the different users.

In one of the most advanced CDMA WLL system currently available, the AirLoop system from Lucent (described in detail in Chapter 12), 16 different PN codes are allowed. The network planner must make sure that the PN codes are assigned to neighboring cells in much the same way that frequencies are assigned in a cluster pattern in a FDMA/TDMA system. That is, there should be at least one cell between two cells that use the same frequency and the same PN code. In practice, that is simple to achieve because sixteen codes are more than adequate. However, it does provide an extra planning stage, of which the network designer must be aware.

7.7 Packet and circuit switching

The types of systems discussed thus far have implicitly assumed circuit switching. That is, when a user starts to make a call, a circuit is established between the user and the network, which is maintained for the duration of the call. The circuit may be an FDMA channel, a timeslot on a TDMA channel, or a CDMA orthogonal code. Whatever, the net effect is that nobody else is able to use that particular resource for the duration of the call.

An alternative to circuit switching is packet switching. In a packet switched system, no permanent connection is established. Instead, the subscriber unit collects data from the user until its buffer is full, then it requests a short slot from the network to transmit the packet of data. The unit then relinquishes the network resources and waits for the buffer to fill again. Packet switching comes in two guises, connection-oriented and nonconnection-oriented. In the case of connection-oriented packet switching, a virtual circuit is established between the transmitter and the receiver, passing through the switching nodes, when the first packet is received. All subsequent packets received for the same destination travel via the same route. Further, they are received in the order in which they are transmitted. In the case of nonconnection-oriented packet switching, each packet is treated as if no previous packet had been sent. Potentially, a packet could be sent via a different route from the previous packet and the packets might not arrive at the receiver in the order they were sent. The receiver then requires a sufficient buffer so it can correctly order the data prior to presenting it to the user.

In outline, circuit switching provides a low and known delay but uses resources inefficiently compared to packet switching. Broadly speaking, circuit switching is suitable for voice, while packet switching is suitable for data. Packet switching is unsuitable for voice because the delays suffered by each packet can be variable, resulting in significant and unwanted voice delay. Packet switching is particularly suitable when the data to be transmitted arrives in short bursts and short delays can be tolerated. That is illustrated by the following example.

Imagine a data source that provided data at the rate of 200 bits every 3 sec and required that the delay on transmission was less than 2 sec (e.g., vehicle location systems provide data in this manner). If the data was transmitted via a circuit switched channel with, say, a data capacity of 9.6 Kbps, a call setup time of 1.5 sec, and a call clear-down time of 1 sec, it would be necessary to maintain the data channel dedicated to that use. If it was attempted to clear down the call between bursts, the signaling required to do so would take so long that the subscriber unit would need to reestablish immediately the channel once it had been cleared down. However, using a packet protocol, with an overhead of 20%, only $200/3 \times 120\%$ bps = 80 bps would be transmitted. That requires less than 1% of the available channel capacity. Packet-mode systems also are ideal for asymmetrical applications where more data is transmitted in one direction than the other (e.g., Internet browsing). Because the uplink and downlink need not be paired, uplink resources are freed for another user who may want to send a large data file in the uplink direction but receive little in the downlink direction.

The ideal radio system probably would include both circuit and packet switched capabilities. Indeed, some modern mobile radio systems, including GSM, are being developed with such dual capabilities. At the time of this writing, all the available WLL systems are circuit switched, but some manufacturers are considering the development of a packet switched capability.

References

[1] Steele, R., ed., *Mobile Radio Communications*, New York: Wiley & Sons, 1992.

[2] Webb, W., and L. Hanzo, *Modern Quadrature Amplitude Modulation*, New York: Wiley & Sons, 1994.

8

TDMA or CDMA?

S INCE ABOUT 1991, there has been substantial debate in the mobile radio journals and more recently in publications addressing WLL about whether CDMA or TDMA is the best access technique. The debate has tended to be distorted and acrimonious because the protagonists are manufacturers of the different access technologies, who want to prove to operators that their equipment is superior. This chapter introduces some of the key aspects to the debate, focusing specifically on WLL.

It should be noted that after six years and the application of most of the finest minds in mobile radio, the debate still has not been resolved. It would be optimistic to expect this chapter to provide all the answers. What it can do is put the discussion on an independent and rational footing and note the key issues and implications for WLL.

The debate has focused mostly on whether CDMA can provide greater capacity for a given spectrum allocation than TDMA. That is clearly a key issue for a WLL operator and is explored in detail in Section 8.1. Other parts of the debate have looked at factors such as ease of cell planning and signal quality and flexibility; each of those factors is addressed in Section 8.2. Finally, Section 8.3 summarizes the implications for WLL, which will be picked up when the separate technologies are examined in Part IV.

8.1 Capacity comparison

8.1.1 TDMA capacity

The capacity of a TDMA system is relatively simple to calculate. The number of channels per megahertz is given by

$$N = \frac{1/_B}{K} \tag{8.1}$$

where B is the bandwidth per channel in megahertz and K is the cluster size. For the purposes of an example, assume that the bandwidth is 8 Kbps and that a cluster size of 3 is achieved through the use of directional antennas. The number of users who can be accommodated, then, is around 40 users/cell/MHz.

8.1.2 CDMA capacity

Direct calculation of the capacity of a CDMA system has proved beyond the capabilities of all the learned scholars who have attempted it. In most approaches, a theoretical analysis is used up to the point where it becomes intractable and then simulation is adopted. When compared with the practical results now being achieved, such an approach looks to have derived results that have proved to be somewhat inaccurate.

Basically, consider N mobiles communicating with a base station. The signal from a particular mobile arrives with a power S. The other mobiles have perfect power control, so their total signal strength is $(N-1)S$. The SIR experienced by the first mobile is

$$SIR = \frac{1}{N-1} \tag{8.2}$$

The SNR is given by the processing gain, G, multiplied by the SIR. Substituting for the SNR and rearranging to determine the number of users, it can be seen that

$$N = \frac{G}{SNR} + 1 \qquad (8.3)$$

which is the basic CDMA capacity equation. It is enhanced as follows, approximating by removing the factor of 1:

$$N = \frac{G}{SIR} \cdot \frac{1}{a} \cdot f \cdot h \cdot p \qquad (8.4)$$

where a is the voice activity factor, f the intercell interference, h the handover loss, and p a factor relating to power-control inefficiency. The calculation of those factors has exercised the cellular community for many years. Fortunately, there is now some practical experience allowing the factors to be examined. What has transpired is that for an equivalent system, cellular CDMA has a capacity of approximately 30% more than cellular TDMA. In such a system, G is 125 and SIR around 5. Thus, (8.4) predicts 25 users/cell/MHz before the other factors, a, f, h, and p are taken into account. In practice, the number of subscribers is around 30% more than TDMA. Because TDMA predicts 40, that suggests that, after the factors a, f, h, and p are taken into account, a capacity of 52 would be achieved; hence, the remaining factors total around 2. Mostly, that gain is achieved through the fact that when a user is speaking there is activity for only 40% of the time.

As far as WLL is concerned, however, the issue is much simpler. The intercell interference is minimal, because directional antennas are used pointing back at the base. There is no handover loss because there is no handover, and the power control is nearly precise. Calculation of the exact effects of those changes is nearly impossible; until systems are widely deployed, the gain of CDMA in a WLL application will not be known exactly. However, it seems likely that in a cellular system $f = 0.66$, $h = 0.85$, and $p = 0.5$; hence, $f \times h \times p = 0.28$. In a WLL system, f might rise to, say, 0.8, $h = 1$, and p might rise to 0.7; thus, $f \times h \times p = 0.56$, approximately doubling the capacity of the cellular system. Overall, it might be expected that a CDMA system would have around twice or even higher the capacity of an equivalent TDMA system when deployed in a WLL configuration. Only a comprehensive trial will demonstrate whether that is correct.

8.1.3   Why the higher capacity with CDMA?

The treatment described in Subsection 8.1.2 is not particularly enlightening. In fact, the reason CDMA achieves a particularly high capacity gain in WLL deployments can be understood intuitively with relatively little effort. Broadly speaking, it has to do with the cluster size. As explained already, in a TDMA system it is not possible to use the same frequency in an adjacent cell because the interference between the two cells will be too high. For that reason, the total available spectrum has to be divided among the cells using different frequencies. That much is well known to those in the cellular industry. CDMA uses the same frequencies in adjacent cells, accepts that interference will reduce the capacity in the first cell, but is able to deploy all the frequencies in all the cells.

The key difference between the WLL and the cellular environments is that WLL subscriber units typically make use of directional antennas. That means they are far less susceptible to interference from other base stations. Similarly, because the antenna is pointing only at a single base station, the base station does not see much interference from subscriber units not in its area. Compared to the cellular environment, in which mobiles have omnidirectional antennas, the interference from adjacent cells is markedly reduced. That is illustrated diagrammatically in Figure 8.1.

In fact, in a well-planned deployment, the interference is reduced to such a level that a TDMA system could, in theory, deploy a cluster size of less than 3, although not as small as 1. In practice, only integral numbers of cluster size are possible, and cluster sizes of 2 cannot be realized well. Hence, TDMA is forced to a cluster size of 3, even though it does not really need it. CDMA does not have that problem. Because the interference from surrounding cells is reduced, the capacity of the main cell increases in proportion. The system still is able to operate in a fully efficient manner. It is that ability to capitalize on the reduced

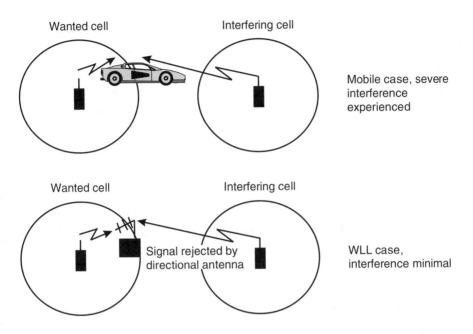

Figure 8.1 Difference in interference when directional antennas are deployed.

interference from adjacent cells that gives CDMA the edge in terms of capacity in a WLL system.

The story does not end there. TDMA has one technique that can overcome this disadvantage, known as *dynamic channel allocation* (DCA). Instead of dividing the frequencies between different cells, each cell has access to all the frequencies. Before transmitting, a cell measures the interference on each frequency and dynamically allocates the channel with the lowest interference. That gives the TDMA system the possibility of using some frequencies that are used in the next cell but for topographical or instantaneous traffic loading reasons are not of a sufficient strength to interrupt transmission. The gains from DCA can be computed theoretically by

$$K_{DCA} = \frac{K_{FCA}^2 + 1}{2K_{FCA}} \tag{8.5}$$

where K_{DCA} and K_{FCA} are the effective cluster sizes for dynamic and fixed channel allocation strategies. Using a fixed cluster size of 3, the dynamic cluster size is 1.66, giving a capacity gain of around 1.8. That reduces the gain of the CDMA system from around 2.5 to a factor of around 1.4, or 40%. Among the different TDMA systems, only DECT and *personal handiphone system* (PHS)–based systems currently offer DCA.

8.1.4 Is CDMA capacity really higher?

Chapter 13 provides estimates for the capacity of a range of WLL systems, using different access technologies. The chapter shows that there are some CDMA systems with a lower predicted capacity than equivalent TDMA systems. Until there are widespread deployments, it is difficult to perform a comprehensive comparison between the different systems.

8.1.5 Should all WLL operators use CDMA?

This section has looked at the capacity of equivalent CDMA and TDMA systems, that is, systems that use the same speech coder, the same channel coder, the same modulation, the same range, and so on. In practice, actual systems are not so similar. Thus, it cannot be guaranteed that a CDMA system always will provide greater capacity than a TDMA system. In particular, the capacity gain of CDMA over TDMA where DCA is used on the TDMA system is relatively small. Also, a number of differences other than capacity are associated with the different systems (those differences are discussed in the next section). In summary, the CDMA systems are not inevitably higher capacity for WLL deployments, but it is highly likely that they will have a higher capacity. Of course, not all operators will need to maximize the system capacity; in rural areas particularly, capacity may be of little concern.

8.2 Comparison of other factors

Besides capacity, a number of other advantages are claimed for CDMA systems. This section examines each of those claims to assess whether they can be supported.

8.2.1 Range

It is claimed that CDMA systems have a greater range than equivalent TDMA systems. Range is related to the path loss and the minimum signal level that the receiver can decode reliably. Path loss is independent of multiple-access methods, so the claim is basically one that CDMA can work with a lower received signal strength than TDMA. Section 8.1 showed that that is true, since the receiver applies a gain, G, to the received signal with a CDMA network but not with a TDMA network. However, to counteract that, TDMA systems need only operate at a higher power level. Such an option is not possible with cellular systems because high power levels rapidly drain mobile batteries; power, however, is rarely a concern for a WLL operator. There seems little reason to support the claim that CDMA systems must have a greater range than TDMA systems.

8.2.2 Sectorization

Sectorization is discussed in detail in Chapter 17. Here, suffice it to say that it is the division of a circular cell into a number of wedge-shaped sectors. It is claimed that if sectorization is performed in a CDMA network, the same frequency can be used in each sector, increasing the capacity of the system by the number of sectors deployed. That claim is correct. It also is claimed that using sectors in a TDMA arrangement does not increase capacity. That also is broadly correct. Fundamentally, when a cell is sectorized, the cell radius remains the same. Hence, the transmitted power remains the same, and the distance required to the next cell using the same frequency also remains the same. However, because there are now more cells within the sterilization radius, more frequencies need to be found to avoid interference. So although the sector is smaller than the cell and thus has to support less traffic, it also has fewer frequencies on which to do that (because the total frequency assignment has been divided by a larger cluster size). That is not the case with CDMA, where using the same frequency in adjacent sectors increases only slightly the interference to neighboring cells and slightly reduces their capacity.

TDMA could achieve a real gain if, instead of making a cell into a number of smaller cells by sectorization, the cell was divided into smaller circular cells, that is, the base stations were distributed around the cell and transmitted on a lower power level. That approach results in similar equipment costs but much higher site rental and backhaul costs; thus, it tends to be avoided except where absolutely necessary.

In summary, the CDMA capacity can be increased by a factor of 2 to 3 by sectorization with only a small increase in cost. The option is not available in TDMA and hence represents an advantage of CDMA.

8.2.3 Frequency planning

When different frequencies need to be assigned to neighboring cells, the network planners have to decide which frequencies to use in which cells. In a CDMA system, where each frequency is used in each cell, no such decision needs to be made. For that reason, it is true that, in general, CDMA does not require frequency planning, although it may require PN code assignment planning. However, that is not a major advantage. Frequency planning can be readily accomplished with today's planning tools and easily adjusted if problems occur. DCA systems do not require frequency planning in any case. Finally, some CDMA WLL systems suggest that frequency planning is performed on a cluster size of 2, for various design reasons, so some frequency planning is required. In summary, frequency planning is not a key issue in the selection process for CDMA.

8.2.4 Operation in unlicensed bands

A number of frequency bands are unlicensed, that is, anyone can use them without having to obtain a license from the regulator. Such bands typically are used by *industrial, scientific, and medical* (ISM) applications, for example, ovens that use radio for heating purposes. Operating WLL systems in those bands has the single attraction that the spectrum is free and that the need to apply for a license is removed. However, the WLL system will suffer unknown and variable interference from uncontrolled sources.

These bands can be used only if the system can tolerate the interference. CDMA implicitly tolerates interference anyway, so in such an environment a CDMA system will work, but with reduced capacity. TDMA systems cannot accommodate such interference, implicitly, but there are techniques that allow them to do so. DCA selects channels according to the interference present on them. Another technique, frequency hopping, moves rapidly from channel to channel, so interference on one channel causes errors only for a short period of time.

CDMA systems cope slightly better with interference because they still use interfered channels but at a lower capacity, whereas TDMA systems use techniques to avoid transmitting on the channels. The capacity of a CDMA system probably is higher, then, in such an environment.

8.2.5 Macrocells versus microcells

The concept of using small cells in high-density areas is discussed in detail in Chapter 17. Suffice it to say here that there are situations where smaller WLL cells are deployed within the coverage area of larger WLL cells. That is a problem for CDMA systems. Subscriber units configured for the larger cell will operate with much higher powers than those configured for the smaller cells. If both cells operated on the same frequency, the capacity of the smaller cell would be near zero, so different frequencies must be used. Because of the wide bandwidth and hence high capacity of a CDMA system, that may be inefficient, in the worst case reducing the equivalent CDMA capacity by a factor of 2. Such reduction does not occur in TDMA, because the cells would be assigned different frequencies in any case.

The actual effect of microcells will vary from network to network, but with good planning, capacity reductions of far less than 2 should be realizable. In summary, this is a problem that, although reducing CDMA's advantage slightly, is unlikely to change the decision to use a CDMA over a TDMA system.

8.2.6 Risk

TDMA systems have been widely deployed around the globe, while CDMA systems are only just starting to emerge. There is a much higher risk with CDMA that equipment will be delayed, will not provide the promised capacity, will prove difficult to frequency plan, and so on. Such risks are continually reducing as experience with CDMA systems grows rapidly. By 1999 the risk probably will have disappeared almost completely. During 1997 and the early part of 1998, operators must decide whether they are prepared to take on such risk and whether there are means whereby it can be ameliorated.

8.2.7 Cost

Everything eventually comes down to cost. At the moment, CDMA system components cost more than TDMA system components. However, because of the higher capacity of CDMA systems, fewer base stations are required, resulting in lower equipment bills and lower site and line-rental costs. How the two facts balance depends on the actual difference in equipment costs and the extent to which the network is capacity limited. Certainly, in a highly capacity-limited situation, CDMA systems should prove less expensive. Other situations are less clear. Part IV looks at the relative system costs when different systems are compared; Part V compares costs in the examination of business cases.

8.2.8 Bandwidth flexibility

CDMA systems can increase the user bandwidth simply by reducing G. TDMA systems also can be bandwidth flexible by assigning more than one TDMA slot per frame to a user. For example, DECT systems can assign between 32 and 552 Kbps dynamically to one user, depending on the load. Thus, both access methods can be made to be approximately equally flexible although manufacturers may not have designed the capability into individual systems.

8.2.9 FH-CDMA versus DS-CDMA

CDMA systems come in many guises. According to textbooks, the type of CDMA discussed so far in this book is known as *direct-sequence CDMA* (DS-CDMA), because the input data is spread by a sequence or a codeword. According to textbooks, there is an alternative known as *frequency-hopped CDMA* (FH-CDMA), in which the bandwidth of the signal is not increased directly but the transmitter jumps from frequency to frequency. Because more than one frequency is used, the effect appears to be to spread over the bandwidth. In practice, only one frequency is being used at one time, so the transmitted bandwidth is not increased, just the spectrum required. The FH is defined as fast when the jumps occur more than once in a bit period and slow otherwise. At the moment, fast FH is restricted to military applications and is not considered further here.

FH has been introduced in the context of TDMA systems that move from channel to channel to avoid interference. By some quirk of history or definition, FH-TDMA and FH-CDMA are identical. It seems far less confusing to use CDMA to mean only DS-CDMA and TDMA to mean TDMA, FH-TDMA, and FH-CDMA—that is the terminology used in this book. Unfortunately, some manufacturers are so keen to be able to place the "CDMA seal of approval" on their equipment that FH-TDMA equipment is labeled as CDMA or, even worse, as FH-CDMA/TDMA. Readers should assume anything labeled with FH is more akin to TDMA than CDMA, although the manufacturer is strictly correct.

8.3 Summary

If the CDMA-versus-TDMA debate were simple, it would have been resolved long ago. In the WLL environment, resolving the debate appears at least slightly simpler than for cellular, and early cellular results provide a useful insight. In summary:

- CDMA systems are likely to have a higher capacity than TDMA systems, ranging from around 1.4 where DCA is available on the TDMA system to as high as 2.5 in some cases. If sectorization is included, capacity can be higher.

- There is unlikely to be any significant range difference between the two systems.

- CDMA does not have an advantage associated with not having to frequency-plan the system.

- CDMA systems perform better in unlicensed bands.

- CDMA systems perform less well where microcells are deployed.

- CDMA may be less costly, but it is more risky.

■ CDMA does not have a significant advantage where bandwidth flexibility is concerned.

Where capacity is key, operators should examine CDMA systems closely, because those systems are likely to provide the greatest capacity for the lowest cost. Otherwise, an operator should select a system according to whether it can provide the required services and other factors such as finance and local maintenance capability. Spending too much time on the CDMA-versus-TDMA debate is an inefficient way to use resources needed to plan and deploy a network.

Part IV

Wireless Radio Technologies

SELECTING THE APPROPRIATE technology often is seen as one of the key decisions facing a WLL operator. Indeed, many operators focus on that decision to the exclusion of all others. It is an extremely important decision, although Part V of this book places the decision in the context of many other equally important decisions that an operator needs to make. There is much confusion in the marketplace and little in the way of independent analysis. This part of the book corrects that situation.

Chapter 9 is an overview of the technologies. Chapters 10, 11, and 12 discuss those technologies: cordless, cellular, and proprietary, respectively. Chapter 13 then provides a summary and conclusions for Part IV.

One important point should be made clear about Part IV. No attempt has been made to list and describe all the available technologies. With the speed that new technologies are being introduced, such an attempt would make the book outdated even before it is published. Instead—and more usefully—the chapters consider examples of each genre of technology and explain the key issues in each general area. That approach will allow readers to understand relatively easily new technologies that subsequently emerge. Also, as a result of the variable availability of information on different systems, some are described in significantly more detail than others.

9

Overview of Technologies

TTEND A CONFERENCE on WLL and typically you will find tens of manufacturers, each with two or more WLL products, making presentations as to which technology you should adopt. Analysts and consultants present single-page diagrams that attempt to summarize which technology is best. Operators state why they chose a particular technology, often contradicting the manufacturers' presentations. Confusion reigns.

At the beginning of 1997, there were some 15 different types of technologies, and some technologies were supported by more than one manufacturer, resulting in around 25 different product offerings. Such a choice is unprecedented in the world of radio!

9.1 Why so many choices?

It is worth understanding why there is such a wide choice. The key reasons are as follows:

- Because no standards have emerged;

- Manufacturers are seeking to adapt equipment from other markets;

- WLL systems are deployed in a range of environments with conflicting needs.

9.1.1 Standards

In the world of cellular radio, three key systems currently are being installed (GSM and its derivatives, CDMA and *digital AMPS*, or D-AMPS). It is hoped that in the future that number will fall to two or even one. The cellular market is larger than the WLL market and has the economies of scale to support more technologies. Only three have emerged because those three have been declared open standards. Equipment from one manufacturer can interwork with equipment from another. That is important to the operators, because they do not feel they are being tied into a particular manufacturer.

Typically, open standards do not emerge by themselves. It rarely is in the interest of a manufacturer to open its technology to other manufacturers. Even when a manufacturer does that, other manufacturers often find that key intellectual property rights are difficult and costly to obtain. Standards work best when a recognized standards committee develops them in an open forum and all interested parties take part. That is what occurred with GSM, which was developed by the *European Telecommunications Standardization Institute* (ETSI)[1]. As a result, most manufacturers produced only a single technology, and operators were faced with a straightforward decision.

The reasons a standard has not emerged for WLL and the actions currently being taken to address that situation are discussed in Section 9.2.

9.1.2 Adapting equipment

As explained in Chapter 4, WLL's history is closely linked with the use of cellular systems to provide telephone lines in some countries. At first,

---

1. It is fondly known to its members as the European Tourism and Sightseeing Institute, because they have had meetings in most European cities.

while it was still far from clear that WLL would be successful, manufacturers sought to reduce their risk by modifying existing equipment rather than developing new bespoke systems. Different manufacturers decided to modify different systems, including their old cellular and cordless systems and their new cellular and cordless systems. Some then started to develop bespoke equipment, leaving their other technologies in the product catalog. As can be imagined, that resulted in a wide range of equipment being available.

9.1.3 Differing needs

It would seem that if one technology were significantly better than another, the number of products on the market would fall rapidly as operators selected the better technology. That clearly has not happened. What is clear, as explained in Chapter 4, is that WLL systems are deployed in a range of environments. Some systems require high range, others high capacity, some certain features, others operation in particular frequency bands. No single technology can meet all those needs. Particular technologies fill particular niches.

This part of the book shows why that is the case and the strengths and weaknesses of the different technologies.

9.2 Standardization activities

Standardization has failed WLL. In many ways, that should not be surprising. Standardization fails the first generation of most products, and WLL has been no exception. The reason is that standardization is time consuming. It takes much longer for a group of manufacturers to agree on a standard than it does for a single manufacturer to design a system in-house. For example, GSM took around eight years to standardize (not including the effort to correct the standard after the initial launch). Qualcomm, working alone, developed its CDMA technology in around three years. So, to produce a standard it is necessary to start early so that the standard is completed around the time that demand for it arrives.

In the early 1980s, first-generation cellular radio (for which there were few standards) was proving successful. It was clear that there eventually would be a demand for a second-generation digital system. ETSI started a standardization program long before the need appeared, resulting in GSM arriving on time (more or less).

WLL just appeared one day, as was recounted in Chapter 4. Nobody had any time to perform any standardization activity before product was available. Once product was available, standardizing it was almost impossible because each manufacturer would want the standard to closely reflect its own system. Just like GSM, ETSI has predicted a demand for a second-generation WLL system, which would be a broadband technology, probably allowing users to have data rates in excess of 2 Mbps if that was what they required. Some standardization work on those systems is underway, and it may be that around 2005 the standard emerges in a timely fashion, allowing the number of WLL technologies to be reduced dramatically. Until that date, it seems highly unlikely that any real standard will emerge.

ETSI has not been totally inactive in the area of first-generation WLL. Although there is little hope of agreeing on the access method and other similar parameters, manufacturers are prepared to agree on limits that the output spectrum should not exceed. Such limits are known as spectral masks and ensure that different systems do not interfere with each other or with other uses of the radio spectrum. A typical spectrum mask is shown in Figure 9.1.

Because FDMA, TDMA, and CDMA systems differ markedly in their spectrum characteristics, ETSI is working on different standards for each type of system. For most readers, those standards will be irrelevant. Some manufacturers may need to change the output stages of their equipment slightly to conform to the standards; otherwise, the standards will have no effect on the market.

Perhaps it is not totally fair to say that there are no WLL standards. Many of the WLL technologies are based on cellular or cordless technologies, which have been standardized. In some cases, the standards include sections that discuss how the equipment should be deployed for WLL (e.g., see the DECT standard). Having standards helps, in that were a WLL operator to select, for example, DECT, as its preferred technology, the fact that DECT is covered by a standard should allow them to

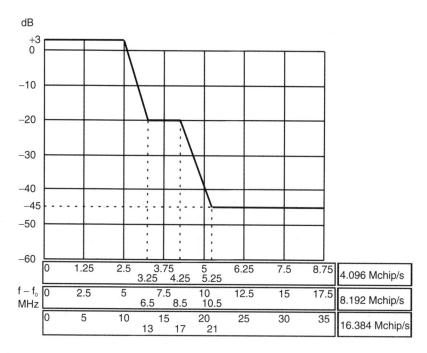

Figure 9.1 Spectrum mask from an ETSI standard (for a CDMA system).

multisource equipment.[2] However, because no WLL-specific standards cover an entire WLL system, there is no guidance from that direction to help an operator make a technology selection.

The lack of standardization is unfortunate; standards undoubtedly provide a number of important benefits to operators and users. One or two standardized WLL systems would provide greater economies of scale, reduced equipment costs, and greater certainty for operators when they must make a technology selection. The lack of standards certainly will slow the progress of WLL and reduce, particularly in early years, the total subscriber numbers. Unfortunately, the lack of standardization cannot be addressed easily at a time when a number of proprietary

---

2. In practice, different manufacturers have slightly modified their DECT equipment to improve performance in a WLL environment. The result is that currently even DECT equipment from different manufacturers will not interwork.

products are available; it will need to wait for the next generation of WLL systems. Standardization, in the same manner that GSM was standardized, would be highly beneficial to the future deployment of WLL systems.

9.3 Segmenting the technologies

Segmentation is an approach much loved by management consultants. In the area of WLL, segmentation is useful because WLL technologies are based on one of three different schools of thought. Understanding that is key to understanding the technologies. Unfortunately, understanding the technologies and determining which is most appropriate for a particular deployment are not the same thing.

Different experts choose slightly different segmentation of the technology market. The general view is that it consists of three different segments, as follows:

- Technologies based on cordless mobile radio standards (e.g., DECT, CT-2, PHS);

- Technologies based on cellular mobile radio standards (e.g., GSM, NMT, IS-95);

- Bespoke or proprietary WLL technologdies (e.g., Nortel's Proximity I, Lucent's Airloop, Tadiran's Multigain).

Further segments could be added. The proprietary technologies could be divided into broadband and narrowband technologies. An extra category could be added alongside broadband for systems designed primarily for video distribution (e.g., MVDS). Such differentiations are important, but they are not sufficiently sizable segments to merit further segmentation of the market.

As a guide, at the start of 1997, the number of cordless, cellular, and proprietary technologies in use, deployment, or trial, is as shown in Figure 9.2.

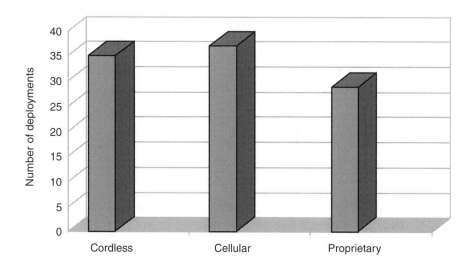

Figure 9.2 Different technologies deployed at the start of 1997.

9.4 Matching the technology to the requirement

As mentioned at the start of this chapter, requirements differ to such a degree that no one technology is perfectly suited to all the requirements. For example:

- Where long range is required, the cordless and proprietary technologies will tend to be inappropriate.

- Where high capacity is required, the cellular technologies will not be appropriate.

- Where good voice quality is essential, analog cellular technologies should be avoided.

- Where ISDN and higher rate services are required, only the proprietary technologies can be used.

- Where interference from existing users of the spectrum is a problem, technologies using CDMA techniques will be most appropriate.

- Where spectrum is congested, proprietary technologies will be required.

- Where speed of deployment is of the essence, cellular technologies will be most appropriate.

- Where limited mobility is required, the cellular and cordless technologies will perform best.

- Where users want a dual-mode cellular/WLL handset, proprietary technologies cannot be adopted.

- For deployments in Europe, non-European cordless and cellular technologies cannot be used.

Other factors could be included. The point of the list is to show how the particular circumstances of the user can have a significant effect on the choice of technology. Before selecting a technology, operators should have a clear understanding of the environment into which they are about to launch their system.

9.5 The important information

In the following chapters on each of the different technologies, it is important that readers keep in mind the key parameters. When selecting a WLL technology, an operator will look for the technology that meets its requirements in terms of services for the lowest possible cost. Hence, the key pieces of information are the services offered and the network cost.

Calculation of the network cost requires an understanding of the range of the technology (so the number of cells for coverage can be calculated), the capacity of the technology (so the number of cells for capacity can be calculated), the architecture of the technology (so the full list of components required can be developed), and the cost of each component. Those key parameters will be provided in Chapters 10 to 13 for each technology, except for cost. Typically, cost is confidential and likely to fall significantly anyway over the coming months.

Calculating capacity is always difficult. For any system, claims often differ as to the capacity that can be achieved, mostly because the capacity can vary, depending on the manner in which the network is deployed. The approach taken here is to determine the bandwidth per channel and the cluster size and them simply divide the bandwidth by the cluster size to arrive at a figure relating to the capacity of each cell. The bandwidth per channel typically is specified by the manufacturer and is not contentious. However, the cluster size is difficult to calculate. Even for a technology such as GSM, where over 200 networks are deployed around the world, estimates of cluster size vary from 3 to 16, a factor of over 500%. With few of the WLL technologies having been deployed in large scales, cluster sizes are likely to be even less well defined.

At this point, many readers might feel that calculation of capacity is a pointless exercise if results can vary by 500%. There is some truth in that. Here, best estimates have been made of the cluster size, using reports of networks deployed, manufacturers' guidelines, and some cluster size theory. That should make the results reasonably comparable, but the capacity figures quoted here should be treated as estimates rather than hard facts.

The following chapters examine each of the technology segments in detail.

10

The Cordless Technologies

C ORDLESS SYSTEMS ARE DESIGNED to replace wired phones and to provide mobility. In the original concept of a cordless system, the user would disconnect a standard wired telephone from the wall jack and replace the phone with a combined cordless home base station and portable handset. The user then could take calls coming into the fixed line using a mobile phone.

A number of important design decisions arose from such origins. First, the range of the phone from the base station needed to be only around 200m, so low power designs were used. Second, the phone and the base station needed to be inexpensive, so the phone design was simple, avoiding complicated speech coders, channel equalizers, and so on. Third, frequency planning was not possible since the users owned the base station as well as the mobile, so the phones needed to seek a low-interference channel whenever they were used. All the technologies detailed in this chapter have those fundamentals very much as a part of their design.

Other applications for cordless phones soon were developed. A key application is the replacement of an office *private access branch exchange* (PABX) and wired phones with a central cordless phone system able to provide mobility and appropriate exchange functions. Systems designed for that application typically use TDMA, to allow the central PABX to use as few amplifiers as possible. DECT is an example of such a system.

The final application for which cordless has been adopted is known under a wide variety of names, most often telepoint. Telepoint is the deployment of cordless base stations in streets and public areas to provide a service that, at first, appears similar to cellular. The key difference is that the base station range is so low that coverage can be provided only in high-density areas, unlike cellular, which provides more ubiquitous coverage. Telepoint systems in general have failed. The United Kingdom launched four networks, all of which failed, as did networks in France, Germany, and several other countries. Typically, the lack of coverage compared to cellular makes the service unattractive. Telepoint has been successful in a limited number of places, namely Singapore, Hong Kong, and more recently Tokyo, all high-density cities where the populace rarely leaves the city areas. In such cities, widespread deployment of cordless base stations can be worthwhile because of the high user density and because relatively good coverage can be provided. The resulting high capacity allows call charges to be lower than cellular, making the service an attractive proposition. It is unlikely that there are many other cities in the world where cordless would be successful in a telepoint application; its main sales will remain as home and office phones.

More recently, cordless technologies have been offered for WLL. From the description given here, it is apparent that the key characteristics of cordless will be low range and high capacity. Some attempts have been made to extend the range using directional antennas, but cordless systems still remain relatively low range. However, the technology is relatively inexpensive, and the need to deploy additional base stations need not incur a cost penalty over other radio systems.

The key cordless technologies of DECT, PHS, and CT-2 are described next. Readers wishing to know more about cordless systems should refer to [1].

10.1 DECT

The *digital European cordless telephone* (DECT) was standardized by ETSI during the early 1990s, initially for use as a wireless office PABX. Only recently has it been suggested for deployment in WLL networks. DECT's key strengths are a good voice quality, the ability to provide a high bandwidth, and the avoidance of frequency planning. Its weaknesses are a low capacity unless deployed with a short range and possible problems as the range increases due to ISI. DECT-based WLL products are available from a range of sources, including Siemens, Ericsson, and Alcatel.

DECT has a further advantage: the availability of unlicensed and standardized spectrum. In most countries, especially in Europe, the frequency band 1.88 to 1.9 GHz has been set aside for DECT. Therefore, operators can obtain spectrum easily, and manufacturers are able to achieve good economies of scale with their equipment. Using the *generic access protocol* (GAP) currently being developed, subscriber equipment from one manufacturer should be able to interwork with network equipment from a different manufacturer, easing concerns over multisourcing.

DECT transmits using TDMA. Each radio channel is 1.728 MHz wide into which a DECT carrier with a data rate of 1.152 Mbps is inserted. Radio channels are spaced 2 MHz apart. Because the data rate is significantly lower than the bandwidth, DECT allows adjacent channels to be used in the same cells, unlike most other radio standards. However, it also makes relatively inefficient use of the bandwidth available. Each 1.152-Mbps bearer is divided into 24 time slots. Nominally, 12 time slots are for base station–to–subscriber transmission, and 12 are for subscriber–to–base station transmission. That can be varied dynamically, however, if there is more information to be transmitted in one direction than the other.

Because of the transmission in both directions on the same frequency, DECT is said to be *time division duplex* (TDD). Literally, the time is divided between the different duplex directions. TDD has a number of advantages, such as simpler RF design. It is important to note that the fact that the same radio channel is used in both directions makes the use of antenna diversity simpler. That issue is explained next in more detail.

The fast-fading pattern changes as the frequency (and hence the wavelength) changes. It is possible for a mobile receiving on one frequency and transmitting on another 45 MHz away to have the downlink in a fade but the uplink transmitting a signal that is received strongly at the base station. One way of overcoming fades is antenna diversity. In such an arrangement, the receiver has two antennas positioned ideally at least a wavelength apart. The hope is that if one antenna is in a fade, the other one is not. By taking the signal from the antenna with the strongest signal during each burst, significant improvements in received error rate can be achieved. That is known as spatial diversity. Unfortunately, the use of diversity antennas on the subscriber unit often is difficult. Because the unit is relatively small, it can be difficult to space the antennas far enough apart, and the additional cost of a second antenna often can make the equipment uncompetitive. TDD provides a solution. When receiving a transmission, the base station decides which of its diversity antennas is receiving the strongest signal. Because the base station will transmit on the same channel back to the subscriber unit, it knows that if it transmits from the antenna that received the strongest signal, then the subscriber unit will receive a good-quality signal. That way, a diversity path is achieved to the subscriber unit without the need for a second antenna in the subscriber unit.

TDD can lead to some confusion in a capacity comparison with a non-TDD system. For a non-TDD system, when capacity per MHz is quoted, it is normal to actually quote capacity on a 2×1 MHz basis, that is, only the spectrum required on the downlink is counted and it is assumed that equal spectrum is available on the uplink. For a fair comparison, TDD systems should have the capacity calculated for a 2-MHz bandwidth. That has been performed throughout this book.

Within each of the 24 slots on the TDMA frame, a 32-Kbps bearer capability is provided. That allows ADPCM speech or a range of data options. Slots can be concatenated to provide up to 552 Kbps per user (requiring 18 of the 24 slots), although that will prevent many other users from accessing the base station at the same time.

DECT also has DCA, which was explained in Chapter 8. DCA allows DECT systems both to exist without the need for frequency planning and to maximize the system capacity in cases where the interference from surrounding cells is relatively low.

The high data rate of DECT of 1.152 Mbps can cause some problems with ISI. DECT does not have an equalizer; hence, any reflected paths with an extra path length of 260m more than the main path can cause problematic ISI. When an omnidirectional antenna is used in an urban environment, the ISI problem tends to limit the range of DECT to under 300m. However, when directional antennas are used at the subscriber equipment, the range can be extended substantially. Figure 10.1 shows that, with a directional antenna with an overall beamwidth of 20 degrees, the subscriber unit can be positioned at least 23 km from the base station before ISI becomes problematic. DECT has a relatively low transmit power, so it is more likely that range would be constrained by inadequate signal strength rather than ISI.

One key advantage of DECT is the simplicity of the system, resulting in relatively inexpensive equipment costs. Because highly complex radio transmission methods and protocols have been avoided, DECT base stations are available at only a fraction of the cost of cellular base stations. Further, with a large number of TDMA slots, 12 on one channel, the RF costs are significantly reduced compared to other systems.

To illustrate a DECT system in more detail, information on the Ericsson DRA1900 system is provided next. The system architecture is

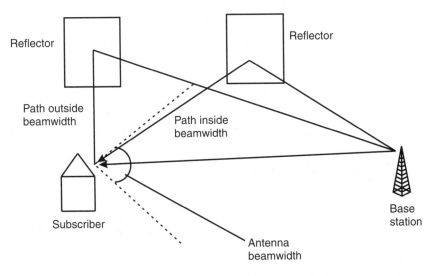

Figure 10.1 Use of the antenna beamwidth to discriminate against long reflected paths.

shown in Figure 10.2. The main components of the system are the *radio node controller* (RNC), the *DECT access node* (DAN), and the *fixed access unit* (FAU). Most of those units are derived from Ericsson PABX products. The RNC is at the heart of the system and acts as a base station controller and a concentrator. It controls calls originating or terminating in the radio network and provides connection to the local PSTN exchange. One RNC can handle 60 simultaneous calls and up to 600 subscribers. One RNC can control up to eight base stations, although in high-traffic areas, the call volume from a single base station will mean that one RNC per base station may be required.

To provide greater capacity, more than one RNC can be deployed. Note that, while subscribers can roam within the area covered by a single RNC, they cannot roam between different RNCs. That is unlikely to be a problem in a WLL environment. Typically, the RNCs will be located at the local exchange site to economize on transmission resources. Because the link between the base station and the RNCs sends voice using 32-Kbps coding, while the link from the RNCs into the network uses 64-Kbps coding, only half the resources are required if the RNCs are deployed in that fashion; for a single RNC, all 60 calls can be accommodated on a single 2-Mbps link.

The base station, or DAN, consists of a small cabinet and two antennas spaced about 2m apart to achieve sufficient spatial diversity. Typically, omnidirectional antennas are used, although directional antennas may be

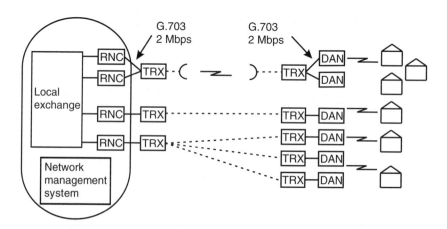

Figure 10.2   Architecture of the Ericsson DECT system.

required where high range is needed. Each cabinet contains six DECT transceivers and a control unit that provides DAN control and multiplexes the call traffic on the industry standard ITU-T G.703/704 2-Mbps connection back to the RNC. The link might be wire, fiber, microwave, or radio. Drop–and–insert multiplexers can be used to connect DANs onto a ring in a bus configuration.

The fixed-access unit is installed at the subscriber's premises. It is composed of a DECT transceiver, a line interface unit, an antenna, a power supply, and a battery backup. The line interface allows an ordinary analog telephone to be connected to the unit. Multiple line units (up to 12) can be provided easily by the connection of more line interfaces to the DECT transceiver.

DECT also has the possibility of deploying repeater units, which may be useful if that is required to provide mobility in a limited area such as an office or shopping mall. The repeater unit works by receiving the data in one time slot during the receive half of the TDD frame and retransmitting it, using a different antenna during the transmit half of the frame. It essentially is a combination of a DECT terminal and a DECT base station. Repeaters are deployed in two ways:

- The coverage-enhancement unit uses multiple directional antennas to extend the range of the DAN, providing coverage in shadowed areas and connecting to the subscriber units in the normal manner. It also can be used to overcome the problem of new buildings being erected after the network was planned.

- The residential fixed repeater unit uses a directional antenna to communicate with the DAN but then reradiates via an omnidirectional antenna, creating a local microcell.

Although repeaters provide some flexibility, they have two significant faults. The first is that they add delay equivalent to the time delay between the receive and transmit parts of the TDD frame, some 10 ms in total. The second is that they are inefficient with resources, requiring twice as much spectrum to carry the call as if a repeater were not used. Nevertheless, in some situations they can be beneficial.

The DECT system also includes a network management component. that manages all the RNCs connected to a local exchange and provides operations and maintenance functionality. Like many *operations-and-maintenance* (O&M) platforms, it runs HP Openview as an operating system, providing a graphical user interface to the network.

For a number of years after the standard was introduced, DECT failed to take off. In recent years, however, penetration has increased substantially, particularly in Germany. It may be that with success in the WLL market DECT will sell substantially around the world. Table 10.1 lists the key parameters of DECT.

10.2 PHS

The PHS is a Japanese standard for use in the 1895 to 1918 MHz frequency band. It has proved highly successful in telepoint applications in Japan, where the subscriber density is high and is generating much interest in Asia-Pacific countries.

Like DECT, it is based on a TDMA-TDD approach. It uses a carrier spacing of 300 kHz, providing four channels, although typically one of those channels will need to be used for control information. PHS uses

Table 10.1
Key Parameters of DECT

Services	
Telephony	Yes, good voice quality
ISDN	Yes
Fax	Yes
Data	Yes, up to 552 Kbps
Videophone	Potentially
Supplementary services	Wide range
Multiple lines	Up to 12
Performance	
Range (radius)	Up to 5 km
Cells per 100 km^2	1.3
Capacity per cell, 2×1 MHz	5.2

32-Kbps ADPCM coding. Also like DECT it uses DCA to optimally select the best frequency. Table 10.2 lists the key parameters of PHS.

10.3 CT-2

CT-2 initially was developed in the United Kingdom as an alternative to analog cordless home phones. Before its development was complete, however, a number of operators suggested that it should be deployed for telepoint. The subsequent failure of telepoint networks in most countries has tended to brand CT-2 as a failed standard. In practice, CT-2 equipment is still manufactured and used for cordless applications around the world and has been suggested for WLL.

CT-2 operates in the 864.1 to 868.1 MHz band using FDMA with channel bandwidths of 100 kHz. Within the FDMA channel, TDD and DCA are used, as with DECT. Each channel can carry 32 Kbps in both directions, providing ADPCM voice coding. CT-2 shares the advantages of diversity gain that DECT achieves with TDD. However, because it

Table 10.2
Key Parameters of PHS

Services	
Telephony	Yes
ISDN	No
Fax	No
Data	Up to 32 Kbps
Videophone	No
Supplementary services	Good
Multiple lines	No
Performance	
Range (radius)	5 km
Cells per 100 km^2	1.3
Capacity per cell, 2×1 MHz	8

cannot concatenate time slots, it does not have the bandwidth flexibility of DECT. It is that limited flexibility that generally makes CT-2 less appropriate than DECT for WLL use.

Table 10.3 lists the key parameters of CT-2.

10.4 Summary of cordless systems

Table 10.4 details the key characteristics of the different cordless systems.

There has been much discussion as to whether DECT or PHS is best. Often the key arguments are lost in the battle between two parties, both of whom have significant commercial interests at stake. It is instructive to examine some of the comparisons to understand better how to select between the two systems.

According to Ericsson, the key differences are as follows:

- DECT has a 30% greater capacity than PHS.

Table 10.3
Key Parameters of CT-2

Services	
Telephony	Yes, good voice quality
ISDN	No
Fax	No
Data	Up to 32 Kbps
Videophone	No
Supplementary services	Limited
Multiple lines	No
Performance	
Range (radius)	5 km
Cells per 100 km^2	1.3
Capacity per cell, 2 × 1 MHz	7

Table 10.4
Comparison of DECT, PHS, and CT-2

Services			
	DECT	PHS	CT-2
Telephony	Yes, good voice quality	Yes	Yes, good voice quality
ISDN	Yes	No	No
Fax	Yes	No	No
Data	Yes, up to 550 Kbps	Up to 32 Kbps	Up to 32 Kbps
Videophone	Potentially	No	No
Supplementary services	Wide range	Limited	Limited
Multiple lines	Up to 12	No	No
Performance			
Range (radius)	Up to 5 km	5 km	5 km
Cells per 100 km^2	1.3	1.3	1.3
Capacity per cell, 2×1 MHz	5.2	8	7

- DECT has a reduced infrastructure cost compared to PHS.
- DECT is easier to plan.
- DECT has a lower tolerance to ISI.
- DECT can provide 64 Kbps and ISDN bearers whereas PHS cannot.
- DECT can provide repeaters.
- DECT can interwork with GSM.
- DECT systems exist better in an uncontrolled environment.

According to Fujitsu, the key differences are as follows:

- PHS has a lower delay.

- PHS has a reduced infrastructure cost.

- PHS has a greater capacity.

- PHS has simpler connection to the PSTN.

- PHS base stations have lower power consumption.

Most of those claims have been discussed. However, it is worth examining the capacity issue in more detail. The starting point is to note that DECT has 12 traffic channels per 2-MHz bandwidth, that is, six per MHz, while PHS has four traffic channels per 300-kHz bandwidth, that is, 13 per MHz. However, PHS requires around 3dB more cochannel protection as a result of its more complex modulation scheme, increasing the cluster size that must be used. The effect of that on the cluster size can be modeled as follows. According to Lee [2], the cluster size is given by

$$K = \sqrt{\tfrac{2}{3} \cdot SIR} \qquad\qquad (10.1)$$

Hence, the cluster size of DECT, with a SIR of around 9 dB, will be 2.3, whereas the cluster size for PHS with a 3-dB greater cluster size will be 3.2. Factoring that into the capacity results in a DECT capacity per cell of 2.6 voice channels/cell/MHz, while PHS achieves four voice channels/cell/MHz. (Remember, the figures quoted in the tables are for 2 × 1 MHz, so they are twice those quoted in this paragraph.)

The question of infrastructure cost is more difficult to evaluate until the actual costs of the different base stations are known. DECT proponents contend that fewer DECT transceivers are required because each transceiver provides 12 channels compared to the 4 of PHS. Ericsson claims that a DECT base station will cost 1.1 times a PHS base station but provide three times the capacity, hence the overall reduction in infrastructure costs. Most likely, the base station costs will be dominated by the economies of scale achieved and not the manufacturing difficulties. In this area, it remains to be seen which technology will achieve the most.

In terms of delay, PHS does have a shorter delay than DECT, but for both systems the delay is so small as to be insignificant (unless DECT repeaters are used). Simplicity of connection to the PSTN and base station power consumption probably are of little relevance.

Perhaps the most interesting differences between DECT and PHS lie in the services each can provide. DECT clearly is much better able to provide high bandwidth and multiple line per subscriber services than PHS. Given that the spectrum efficiency is relatively similar and costs probably are not too dissimilar, it is that improved functionality that gives DECT the edge over PHS.

As a final point of note, the PHS spectrum allocation is not assured in the *Central European Post and Telecommunications* (CEPT) countries. Hence, it may prove more difficult to deploy PHS than DECT from a spectrum allocation viewpoint in all but the Asia-Pacific countries.

References

[1] Tuttleby, W., ed., *Cordless Telecommunications World-Wide*, Berlin, Germany: Springer-Verlag, 1996

[2] Lee, W. C. Y., "Spectrum Efficiency in Cellular," *IEEE Trans. Vehicular Tech.*, Vol. 38, No. 2, May 1989, pp. 69–75.

11

The Cellular Technologies

M OST LIKELY, few readers will not have a cellular phone, so a description here of the service offered is largely unnecessary. Cellular phones have been immensely successful, first in the mid-to late 1980s with analog systems and then in the early to mid-1990s with the second-generation digital systems. The key strengths of cellular include good coverage, excellent immunity to channel errors, excellent economies of scale as a result of high sales, and advanced and fully featured phones. Given the use of cellular in some of the first WLL applications, it is only natural that cellular should find itself a key WLL technology.

Cellular technologies also might seem the natural choice where cellular operators have been given WLL licenses (although the provision of dual licenses has not been a widespread occurrence). The use of cellular technology for WLL would bring benefits in terms of economies in operations and maintenance. However, mobile operators typically will not want to use their mobile spectrum for fixed applications and for that reason may select a different technology in a different band.

When applied to WLL, cellular brings the advantage of immediate availability, inexpensive subscriber equipment, and proven systems. It has good range and can offer mobility if required. However, it also brings a host of problems, including relatively poor speech quality (although this is improving); expensive network infrastructure configured to handle mobility, handover, and so on, which is no longer required; and a need to use spectrum dedicated to cellular services, which tends to be expensive.

There are a number of analog cellular systems, but generally they are so similar that it is sufficient to describe one typical system to characterize them all. Broadly speaking, there are three digital systems: GSM, IS-95, and D-AMPS. DCS1800 and PCS1900 simply are variants of GSM operating at different frequencies. D-AMPS and GSM both are TDMA based, and it is sufficient to describe just one to characterize the systems. IS-95 is CDMA based and must be described in detail to understand its key attributes.

11.1 Analog cellular

Analog cellular has been deployed widely throughout most of the first world countries since the mid-1980s. It slowly is being withdrawn as digital cellular networks are rolled out. Its key attraction for WLL lies in the fact that it is readily available and relatively inexpensive. Its key problems are that it lacks facilities and has poor voice quality. The wide deployment of analog cellular is expected to decrease in the future as digital and proprietary equipment becomes more readily available. As far as WLL operations are concerned, all the analog cellular systems are more or less identical. The *total access communications system* (TACS) is described here as an example.

A TACS network consists of a number of base stations connected directly to a switch. Switches are interconnected and are connected into the PSTN. The TACS system allows for registration, tracking, and handoff, but because those features are irrelevant for WLL, they are not discussed further here.

TACS operates in the 900-MHz band using a 25-kHz channel spacing and a 45-MHz duplex spacing. Each TACS network also requires at least 21 channels to be reserved for control purposes. The control channels provide paging information, allow random access, and provide the mobiles with information on the system status.

Speech is carried within the radio channel using analog FM with a peak deviation of 9.5 kHz. That deviation is relatively high, resulting in significant adjacent channel interference. For that reason, adjacent channels cannot be used in the same cell.

One major problem of TACS is that when a mobile is in a voice call, signaling also must be sent over the same channel. When a signaling message is being sent, the radio momentarily mutes the audio path, which results in occasional loss of a syllable of voice. The effect is a slight, but noticeable, degradation in the speech quality.

Some of the key parameters of TACS are provided in Table 11.1. In calculating the capacity, a cluster size of 12 was assumed, which, if anything, results in an optimistic assessment.

Table 11.1
Key Parameters of TACS

Services	
Telephony	Relatively poor quality
ISDN	No
Fax	No
Data	Limited, perhaps 1.2 Kbps
Videophone	No
Supplementary services	Limited range
Multiple lines	No
Performance	
Range (radius)	Up to 35 km where topography allows
Cells per 100 km^2	0.026
Capacity per cell, 2 × 1 MHz	3.3 voice calls

11.2   Digital cellular

There are two key digital cellular systems, those based on TDMA technology and those based on CDMA technology. Examples of both are provided in this section.

11.2.1   GSM / DCS1800

GSM is a TDMA digital cellular system deployed in more than 120 countries around the world. It is the nearest piece of equipment yet to become a global mobile standard. It has been extremely successful as a result of good technical design and the fact that it is an open standard allowing operators to multisource equipment both on the network and mobile side. For a full description of GSM, see [1].

A GSM network consists of base stations connected to *base station controllers* (BSCs), which are connected in turn to switches. The switches in the network are interconnected, and there is a single O&M and billing platform located in the network.

GSM divides the spectrum into 200-kHz TDMA channels. Each channel is modulated with the use of *Gaussian minimum shift keying* (GMSK), resulting in significant adjacent channel interference. As with TACS, it is not possible to use the adjacent channel in the same cell. The 200-kHz channels are clustered with a cluster size of around 9. Some manufacturers claim that much smaller cluster sizes can be achieved if frequency hopping is employed. At present, frequency hopping is little used because some early mobiles did not implement it correctly, resulting in users being dropped from the network when hopping is attempted. However, in a WLL network it could be guaranteed that all subscriber units could hop, and the directional antennas would help reduce the cluster size. Hence, the assumption is made here that a cluster size of 4 could be realized in a WLL deployment.

Each 200-kHz channel is divided into 8 full-rate voice channels or 16 half-rate voice channels. The half-rate channels use a lower rate coder, which generally is perceived to be of a lower speech quality. Nevertheless, the important capacity savings mean that some cellular operators are deploying half-rate systems. Each voice channel has an associated control channel. This channel allows information such as the frequencies of surrounding cells to be passed to the mobile and for the mobile to return measurement information without causing the break in speech that would occur in TACS. At least one voice channel per cell needs to be set aside as a control channel, providing paging and cell broadcast information.

GSM employs 13-Kbps speech coding using a complex RPE-LTP coder. That provides speech quality that is better than analog cellular but typically inferior to ADPCM coders. The GSM speech coder is soon to be enhanced with the so-called enhanced full-rate coder, which should provide a significant improvement in voice quality, possibly to a level equal to ADPCM.

Each voice channel alternatively can be used as a data channel. GSM provides a range of data rates depending on the error rate that can be tolerated. The services available are described in Table 11.2.

GSM also can transport group 3 fax information transparently over the air interface. GSM currently is being enhanced to provide packet data and higher speed data. That would allow data rates of up to 64 Kbps at a 0.3% error rate and more efficient use of the air interface resources. Table 11.3 lists the key parameters of GSM.

11.2.2 IS-95

The IS-95 standard for cellular communications is based heavily on the CDMA technology designed by the American company Qualcomm. The standard has gained some success in the United States and Asia-Pacific regions and is now entering commercial deployment. Its proponents have claimed that its major advantage is a significant capacity increase over other cellular systems. That increase has proved impossible to calculate

Table 11.2
GSM Available Services

Data Rate	Channel Type	Error Rate
9.6 Kbps	Full	0.3%
4.8 Kbps	Full	0.01%
	Half	0.3%
2.4 Kbps	Full	0.01%
	Half	0.001%

Table 11.3
Key Parameters of GSM

Services	
Telephony	Yes, with digital voice quality slightly inferior to wireline
ISDN	No, but 64-Kbps support may be available in 1999
Fax	Yes
Data	Yes, up to 9.6 Kbps
Videophone	No
Supplementary services	Wide range
Multiple lines	No
Performance	
Range (radius)	Up to 30 km where topography allows
Cells per 100 km^2	0.035
Capacity per cell, 2 × 1 MHz	Full rate: 10 voice channels; half rate: 20 voice channels

or simulate, but now that the first results are available from deployed networks, it can be stated with some confidence.

The IS-95 standard specifies only the air interface, allowing equipment manufacturers to design the network in whatever manner they deem appropriate. In practice, many manufacturers have adopted the GSM design principles; hence, base stations connect back to BSCs and then to switches.

The system uses single cell clusters, in which each base station transmits its CDMA signal on the same carrier. The carrier bandwidth is 1.228 MHz, and there is a duplex spacing of 45 MHz.

The downlink transmission consists of a permanent pilot tone and a number of radio channels. The pilot tone is used by the mobile to estimate the path loss, so as to set power control initially, and to acquire synchronization to the codeword generator. Other channels are set aside for paging and other downlink information. Speech is encoded using a variable-rate codec, where the bit rate depends on the talker activity. That reduces interference to other users of the CDMA channel. Channel coding then is applied to generate a transmitted bit rate of 19.2 Kbps, which then is spread by a 64-chip Walsh code sequence to generate the 1.228-Mbps transmitted waveform.

The uplink transmission is slightly different. Speech is generated in the same manner, but higher rate coding is used to give a bit stream of 28.8 Kbps. The interleaved bits then are grouped into 6-bit symbols, and each symbol addresses a lookup table containing a particular Walsh code. That spreads the signal to 307.2 kchips/s. Because other mobiles could select the same Walsh code, further spreading is required. Each mobile generates a unique PN sequence at 1.228 Mbps, which is used to spread the data to the full transmitted bandwidth.

Calculation of the capacity of a CDMA system is extremely difficult. As explained in Chapter 8, some approximations can be made. However, perhaps more useful are the results from real deployments where up to 15 voice channels per carrier were employed. This was achieved with the 13-Kbps voice codec described next. CDMA has a key advantage over TDMA in that when a new sector is added, the capacity is increased by an extra 15 channels. In TDMA, the addition of new sectors provides only limited capacity gains. That makes it difficult to perform a strict comparison of the different technologies. Here, a single sector per cell has been assumed, but it is noted that the CDMA system can be enhanced more readily than the TDMA system.

When IS-95 was first standardized, it used an 8-Kbps proprietary voice codec designed by Qualcomm. However, there was some concern about the voice quality that could be supported with this codec. Thus, a second codec was introduced, similar to the GSM codec and working at 13 Kbps. The majority of network operators seem to be selecting this 13-Kbps codec, and it seems likely that it would be used for WLL. For that reason, capacity calculations here have been performed for the 13-Kbps codec.

The IS-95 standard is relatively poor at supporting fax and data. Both can be carried with modem support up to around 9.6 Kbps, but the range of services provided by GSM is not available. It may be that those services will be introduced in the future. Table 11.4 lists the key parameters of IS-95.

Table 11.4

Key Parameters of IS-95

Services	
Telephony	Yes, with voice quality similar to or slightly better than GSM
ISDN	No
Fax	Yes, using a modem
Data	Up to 9.6 Kbps
Videophone	No
Supplementary services	Limited range
Multiple lines	No
Performance	
Range (radius)	30 km under good topographic conditions
Cells per 100 km^2	0.035
Capacity per cell, 2×1 MHz	12 per sector

11.3 Future cellular systems

Substantial work is being carried out on the next generation of mobile radio systems, which are expected to appear between 2002 and 2005. Although no information is available on whether those systems will be designed such that they are applicable to WLL, it seems likely that the technology eventually will be suggested for WLL. Very little is known about these systems, including their data rates, facilities, and so on, so at this stage it is not possible to understand whether they will be attractive for WLL operation. This section briefly describes the ideal third-generation mobile systems.

First, before the start of any explanation of the third-generation ideal, it is important to understand certain terminology. In Europe the next generation has been named the *Universal Mobile Telecommunications System* (UMTS); at a worldwide level, it has been termed the *future public land mobile telecommunications system* (FPLMTS).

The vision of UMTS, which is shared to a large extent by those outside Europe, has been articulated in many slightly differing ways but can be summarized as "communication to everyone everywhere." Communication in that instance might include the provision of information. It will be a system that everyone will use regardless of whether they are in the office, at home, or on an airplane. To achieve that goal, the system essentially must provide the following features:

- An extensive feature set to appeal to all the current disparate users of different types of radio system such as:

  Cellular users, who require high voice quality and good coverage;

  Users who currently deploy their own private systems (often known as PMR users), who require group and broadcast calls, rapid channel access, and low cost;

  Paging users, who require a small terminal and good coverage;

  Cordless users, who require excellent communications with high data rates when in the office or home;

  Satellite users, who require truly worldwide coverage;

  Users in airplanes, who currently have limited system availability;

  Data users, whose requirements range from telemetry to remote computer network access.

- Ubiquitous coverage with a wide range of cells such as:

  Satellite cells covering whole countries;

  Macrocells covering a radius of up to 30 km;

  Minicells covering up to around 3 km;

  Microcells covering a few streets;

  Picocells covering an office, a train, an airplane, and so on.

Major players in this area postulate that that will be achieved through a system with the following characteristics:

- An adaptive air interface such that access methods such as CDMA, TDMA and FDMA can be selected as appropriate with bandwidths dependent on the service required and a data rate of up to 2 Mbps available in some locations;

- Mobiles that can have their "operating system" downloaded to allow for network evolution;

- An architecture based on intelligent network principles.

The system also must integrate seamlessly with the fixed network such that users receive nearly identical services whether they are using fixed or mobile phones. Current predictions as to the time scales of third-generation systems vary slightly, but typically the following assumptions are made:

- Standardization completed, 1999;

- First product available, 2002;

- Product widely available, 2005.

One of the prerequisites of the third-generation system was that it was agreed worldwide as a single global system, achieving the design aim of international roaming. Recently, however, that goal has started to look like an increasingly unlikely outcome. There are three main key players in third-generation work: Europe, the United States, and Japan. Europe would like to see GSM evolve to become the third-generation system. The Japanese have a different agenda: they have failed to make an impression on the world scene with their first- and second-generation systems and want to make sure they do not fail with the third-generation system. They plan to do that by launching a new system as early as possible, based on something quite different from GSM. Finally, the United States has the view that as little as possible should be standardized, because standards prevent innovation and consumer choice and because standards making often is not performed by the most appropriate bodies. Some U.S. manufacturers have gotten together and are promoting a wider-band version of CDMA, one that may become another contender for third-generation systems. The pressures from these different entities have risen to the extent that most observers expect third-generation systems to include a variety of standards and for third-generation phones to be multistandard so they can work in whatever country they are used.

Whatever the route, by the year 2005 it is expected that there will be a new generation of cellular systems. The systems will have a much wider range of capabilities, support higher data rates, and be able to communicate with satellites when out of the range of cellular systems. It is likely that they also will be used for some WLL deployments, perhaps by the year 2005.

11.4 Summary of cellular systems

The features of the different cellular systems discussed in this section are summarized in Table 11.5.

Table 11.5
Comparison of TACS, GSM, and IS-95

Services			
	TACS	GSM	IS-95
Telephony	Relatively poor quality	Yes, slightly inferior quality	Yes, slightly inferior quality
ISDN	No	No	No
Fax	No	Yes	Yes
Data	Limited, perhaps 1.2 Kbps	Yes, up to 9.6 Kbps	Up to 9.6 Kbps
Videophone	No	No	No
Supplementary services	Limited range	Wide range	Wide range
Centrex	No	No	No
Multiple lines	No	No	No
Performance			
Range (radius)	Up to 35 km	Up to 30 km	30 km
Cells per 100 km^2	0.026	0.035	0.035
Capacity per cell, 2×1 MHz	3.3 voice calls	10 voice channels	12 voice channels per sector

As Table 11.5 shows, the digital systems have clear advantages in terms of services offered and capacity. No costs are provided here, but digital prices are falling so fast that digital systems, especially GSM, can now be less expensive than analog systems. It seems likely that in the future digital cellular will be chosen in preference to analog cellular in new WLL deployments. Of the digital systems, there seems little difference between GSM and IS-95. But, as explained in Chapter 8, a number of factors are associated with WLL deployments that might lift the capacity of CDMA networks by a factor as high as 2, resulting in CDMA potentially having up to 2.5 times the capacity of GSM in some networks. Thus, if capacity is key and a cellular technology is required, CDMA probably should be deployed. However, with its lower economies of scale and limited multisourcing, CDMA is more expensive than GSM on a cost-per-base-station basis. For that reason, a full analysis should balance the fewer base stations required for CDMA against the increased cost for each base station.

The number of networks in which different cellular technologies are deployed as of the start of 1997 is shown in Figure 11.1.

RAS is a system developed by Ericsson based on the analog NMT standard. WiLL is a system by Motorola based on the AMPS U.S. analog standard, and CDMA WiLL is a Motorola system based on the IS-95

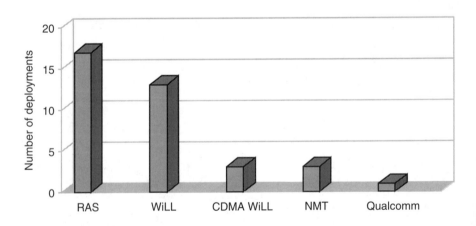

Figure 11.1 Deployment of cellular technologies in WLL networks at the beginning of 1997.

standard. NMT is a system by Nokia based on the analog NMT standards, while Qualcomm is offering a system based on the IS-95 CDMA standard.

To date, of the cellular systems sold, analog cellular comprises around 80% of all the WLL networks using technology based on cellular standards. Generally speaking, that is because the IS-95 equipment is only just becoming available, and the GSM manufacturers have been working at capacity to meet demand from the cellular operators and thus have been disinclined to target the WLL marketplace. The situation has changed, and it is predicted that few WLL networks based on analog cellular technology will be announced in the coming years.

Reference

[1] Mouly, M., and M-B. Pautet, *The GSM System for Mobile Communications*, published by the authors, ISBN 2-9507190-0-7.

12

The Proprietary Technologies

PROPRIETARY TECHNOLOGIES are defined as those designed specifically for WLL, so it might be imagined that they would be superior to the cordless and cellular technologies. Technically, that probably is true, but just as custom-made suits are not for all, bespoke WLL technologies are unlikely to be found in all networks.

The proprietary technologies generally share the following characteristics not available in the other technologies:

- They all offer a wide range of services, including ISDN support.

- They operate in frequency bands above 2 GHz.

- They provide excellent voice quality.

Proprietary technologies typically are more expensive than cellular and cordless. However, in some environments the advantages listed above are so overwhelming that the cost increment is not relevant.

The wide range of proprietary technologies is increasing monthly. No attempt is made here to explain all the technologies. Instead, those technologies that either are particularly popular or have some particularly unusual characteristics are explained. To form some sort of grouping, technologies are split as to whether they are TDMA or CDMA. In addition, a short section is devoted to the MVDS, a form of WLL concentrating on broadcast. In some cases, because of confidentiality issues, only relatively small amounts of information could be gathered on certain systems. Also, many of the technologies are not in full commercial operation, so key figures such as capacity are estimates.

12.1 The TDMA technologies

12.1.1 Nortel Proximity I

Nortel make a range of WLL systems under the banner Proximity. All the systems with the exceptions of Proximity I are based on cordless or cellular technologies. Proximity I is a TDMA system developed in conjunction with the United Kingdom WLL operator Ionica, one of the first operators in the world to deploy a proprietary WLL system. At the start of 1997, Ionica had achieved coverage of over 700,000 homes, and 15,000 customers had been connected. Those numbers are expected to increase rapidly in coming years.

Proximity I offers a wide range of services, including 64-Kbps voice and data links and a second-line capability. Subscriber units link to base stations over the air interface, and base stations then are connected directly back to a PSTN switch. Each subscriber unit provides two lines.

The system uses TDMA channels of bandwidth 307 kHz in a cluster size of 3, transmitting 512 Kbps over the channel using *quadrature phase shift keying* (QPSK) modulation. Up to 54 TDMA bearers can be accommodated in the 3.4-3.6 GHz assignment using *frequency division duplex* (FDD) with a maximum of 18 channels on any given base station. Each TDMA channel can support ten 32-Kbps data channels. DCA is not provided, but it is relatively easy to reconfigure the frequency assignment from the operations and maintenance center. Table 12.1 lists the system's key parameters.

Table 12.1
Key Parameters of Nortel Proximity I

Services	
Telephony	Yes, high-quality voice
ISDN	For future release
Fax	Yes
Data	Yes, up to 64 Kbps
Videophone	No
Supplementary services	Wide range
Multiple lines	Two lines
Performance	
Range (radius)	Up to 15 km in rural areas
Cells per 100 km^2	0.14
Capacity per cell, 2×1 MHz	11

12.1.2 Tadiran Multigain

Tadiran will not be happy to see its system classified as TDMA rather than CDMA. They actually market it as an FH-CDMA/TDMA system. However, according to the terminology adopted in this book (see Chapter 7), their system is in fact an FH-TDMA system. Given the relatively small size and exposure of Tadiran, they have been remarkably successful in deploying their system to date, and a number of operational networks around the world are using the Tadiran technology.

In the Tadiran system, users transmit in a given TDMA slot. However, the actual frequency in which they transmit changes from burst to burst, where a burst lasts 2 ms (hence, there are 500 hops/s). In a given cell, no two users transmit on the same frequency at the same time. However, users may transmit on the same frequency in adjacent cells. By employing different hopping sequences in adjacent cells, if a collision does occur it will be only for a single burst. Error correction and interleaving largely can overcome the effect of such a collision. The system has the advantages of the simplicity of a TDMA system coupled with some of the "interference-sharing" properties that make CDMA spectrally efficient. Tadiran claims that a cluster size of 1.25 can be achieved by that approach when directional antennas are deployed. In practice, that seems somewhat optimistic, and it might be expected that a cluster size of 2 would be more realistic.

The system uses a voice coder of 32 Kbps. It employs TDD, in which both the uplink and the downlink are transmitted on the same frequency but at different times. Each 1×1 MHz channel supports eight voice channels. Hence, 16 voice channels per 2×1 MHz can be supported before the cluster effect is taken into account, and assuming a cluster size of 2, around 8 voice channels/cell/2×1 MHz. Table 12.2 lists the key parameters of the Tadiran Multigain system.

12.2 The CDMA technologies

12.2.1 DSC Airspan

The DSC Airspan system was developed in conjunction with BT, which is using the system for rural access at 2 GHz. The system provides

Table 12.2
Key Parameters of Tadiran Multigain

Services	
Telephony	Yes, good voice quality
ISDN	For future release
Fax	Yes
Data	Yes, up to 32 Kbps
Videophone	No
Supplementary services	Yes
Multiple lines	1 to 4 lines
Performance	
Range (radius)	Up to 6 km in rural areas
Cells per 100 km^2	0.88
Capacity per cell, 2×1 MHz	8

64-Kbps voice channels and support for up to 144 Kbps ISDN services. DSC claims a cluster size of between 1 and 3, depending on the environment.

Voice currently is provided using 64-Kbps PCM. ADPCM at 32 Kbps likely will be available before the end of 1997. The system currently provides 2B + D ISDN per subscriber or, alternatively, 2 × 64 Kbps data channels. Up to 6 × 64 Kbps data channels can be achieved by combining three subscriber units. By mid-1998, units capable of handling 384 Kbps in the downlink direction coupled with 64 Kbps in the uplink are expected to be available. Such units would be targeted mostly at Internet use and would take advantage of the fact that the system capacity is greater in the downlink direction than in the uplink.

Radio channels are 3.5 MHz wide. Each 3.5-MHz channel provides up to fifteen 160-Kbps radio bearers. With the current deployment, each 160-Kbps bearer could provide two voice channels but only to the same house. By the end of 1997, each bearer will be able to provide four 32-Kbps voice channels, each to a different house. Cluster sizes of 2 are possible using sectored cells.

Note that, at the moment, the system is nonblocking. That means a single 160-Kbps bearer is assigned permanently to each subscriber, significantly increasing the network resources required. Standard trunking-based systems are expected to be available during 1997.

Table 12.3 lists the key parameters of the DSC Airspan system.

Table 12.3
Key Parameters of DSC Airspan

Services	
Telephony	Yes
ISDN	Yes, up to 144 Kbps
Fax	Yes
Data	Yes, up to 128 Kbps
Videophone	No
Supplementary services	Wide range
Multiple lines	Yes

Table 12.3 (continued)

Performance	
Range (radius)	5 km
Cells per 100 km^2	1.3
Capacity per cell, 2 × 1 MHz	8.5

12.2.2 Lucent Airloop

The Lucent Airloop technology is a CDMA-based system developed for a wide range of customers. It operates mainly in the 3.4-GHz band using 5-MHz wide channels, each supporting 115 16-Kbps channels. To support 32-Kbps ADPCM, two channels are used simultaneously. The spreading code is 4096 Kbps; thus, for a 16-Kbps data rate, a spreading factor of 256 is used. The top-level network architecture is shown in Figure 12.1.

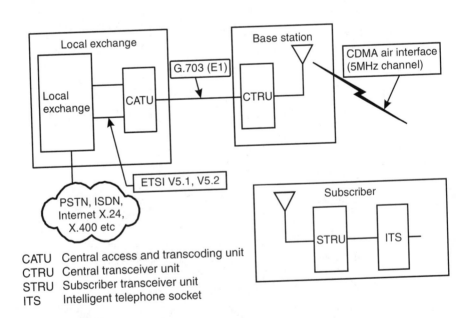

CATU   Central access and transcoding unit
CTRU   Central transceiver unit
STRU   Subscriber transceiver unit
ITS   Intelligent telephone socket

Figure 12.1   Radio access architecture.

The system employs a network of *radio base stations* (RBSs) to provide coverage of the intended service area. The main functional blocks of the network are described next.

12.2.2.1 Central Office

The central office contains digital switching and network routing facilities required to connect the radio network to ISDN and the Internet.

12.2.2.2 Central access and transcoding unit

The *central access and transcoding unit* (CATU) is colocated with the switch and performs the following functions:

- It controls the allocation of radio resources and ensures that the allocation is appropriate to the service being provided, for example, 64-Kbps digital, 32-Kbps speech, ISDN.

- For speech services, it provides transcoding between various speech-coding rates and the switched 64-Kbps PCM.

12.2.2.3 Central transceiver unit

The *central transceiver unit* (CTU) is located at the RBS. A single CTRU provides a single CDMA radio channel. The CTU performs the following main functions:

- It provides the CDMA air interface.

- It transfers ISDN and *plain old telephony* (POTS) signaling information transparently between the air interface and the CATU.

The connection between the CATU and the CTRU requires a 2-Mbps capability and is normally made using direct connection or point-to-point microwave links.

12.2.2.4 Network interface unit

The *network interface unit* (NIU) connects the subscribers to the radio network. The NIU comprises two functional blocks, the *intelligent telephone socket* (ITS) and the *subscriber transceiver unit* (STRU).

- The ITS provides the point of connection to the subscriber's terminal equipment, for example, PABX, telephone, or LAN. The ITS is available in dual-line or multiline configuration.

- The STRU is located on the outside of the subscriber's building and consists of an integrated antenna and radio transceiver. The STRU provides the interface between the ITS and the CDMA air interface. The STRU is connected to the ITS by a standard four-wire telephone or data networking cable.

The number of subscriber connections supported by each NIU is determined by the type of service being provided by the connection. The basic NIU connection provides a single ISDN (2B + D) connection, effectively giving two unrestricted 64-Kbps channels. The same unit also can be configured as either two or eight individual POTS lines using ADPCM and *Code excited lireco predictor* (CELP) speech coding, respectively.

The modulation technique employed first takes each 16-Kbps channel and adds error-correction coding to reach 32 Kbps. It then uses Walsh spreading with a spreading factor of 128 to reach the transmitted data rate of 4 Mbps. Finally, it multiplies that by one of a set of 16 PN code sequences also at 4 Mbps, which does not change the output data rate but provides for interference from adjacent cells, as described in Section 7.6. During design of the network, each cell must have a PN code sequence number assigned to it such that neighboring cells do not have the same number.

Table 12.4 lists the key parameters of the Lucent Airloop system.

12.3 Other technologies

12.3.1 Phoenix

The Phoenix system is described briefly here because it is somewhat different from the other systems. Phoenix manufactures flexible switches based on computing platforms. It has designed a number of software packages that allow the switch to connect to base stations from a range

Table 12.4
Key Parameters of Lucent Airloop

Services	
Telephony	Yes, good quality
ISDN	Yes
Fax	Yes
Data	Up to 128 Kbps
Videophone	No
Supplementary services	Good
Multiple lines	Yes, 2 or 8 line units
Performance	
Range (radius)	4 km
Cells per 100 km^2	2
Capacity per cell, 2 × 1 MHz	5.75

of different technologies, basically most analog and digital cellular standards. That allows a WLL operator to buy a smaller and more flexible switch than might otherwise have been required. For that reason, the Phoenix system probably is most suitable for rural areas.

12.3.2 Microwave video distribution systems

A number of systems have been proposed and implemented that use WLL techniques to deliver broadcast TV. Such systems fall into a niche somewhere between terrestrial TV broadcasting and WLL telephony delivery. They offer advantages over terrestrial TV broadcasting in that they can provide many more channels and may offer advantages over the other WLL systems discussed here in their ability to deliver high-bandwidth services. Broadly speaking, the only difference between WLL and microwave distribution is that the latter tends to be transmitted at much higher frequencies, such as 40 GHz, where significantly more bandwidth is available and hence wider bandwidth services can be offered. However, such higher frequencies result in lower propagation distances and more costly equipment. Further, rain fading can be a significant problem in some countries at such frequencies, making reception unreliable.

The first microwave distribution systems were implemented in the United States, where they were called *microwave multipoint distribution systems* (MMDS). The systems tended to operate at around 2.5 GHz and provided analog TV transmission to communities. They were entirely broadcast systems, and no return path capability was provided. Many such systems are still in existence between 2.5 and 3 GHz, but with the introduction of digital broadcasting they will become increasingly outdated.

After MMDS, digital distribution systems operating at around 29 GHz were introduced in the United States and the Asia-Pacific countries. The systems, known as *local multipoint distribution systems* (LMDS), can provide many more channels with a higher quality but lower range. Similar initiatives in the United States and Europe have lead to the system operating at 40 GHz, MVDS.

Early in the development process of MVDS, it became apparent that to provide competition to cable and to maximize the revenues that could be achieved, a return path from the home to the network would be required. That allows voice and limited data enabling, for example, selection of video films. Systems capable of providing such a return path are now in a trial stage. The return path can provide around 20 Kbps of data and adds around $200 to the cost of the subscriber equipment.

MVDS systems still are relatively immature, so it is difficult to provide significant amounts of information on particular products. Table 12.5 summarizes the likely capabilities of MVDS.

Table 12.5
Key Parameters of MVDS

Services	
Telephony	Potentially acceptable quality
ISDN	No
Fax	No
Data	50 Mbps to the home, 20 Kbps to the network
Videophone	No

Table 12.5 (continued)

Services	
Supplementary services	Poor
Multiple lines	No
Performance	
Range (radius)	1 km
Cells per 100 km^2	31
Capacity per cell, 2 × 1 MHz	Unlimited to the home, relatively low to the network

It can be seen that in terms of telephony provision, MVDS is inferior to other WLL systems. However, when telephony is viewed as a service offered on the back of video distribution, it looks more attractive. It is too early to say whether the economics of MVDS will allow the telephony component to be sufficiently cheap that users will accept its relative shortcomings. However, for many users, telephony is a critical service that they will not compromise to realize some savings. Hence, at least for the next few years, it is unlikely that microwave distribution systems will provide an acceptable WLL offering. Instead, they provide a wireless alternative for the cable operator.

12.3.3 The broadband technologies

A number of new technologies currently under development at the moment are aimed at providing broadband telecommunications services. Unfortunately, broadband remains a loosely defined term. In the area of WLL, it typically refers to bandwidths greater than 2 Mbps per subscriber, but that is not definitive. Broadband systems have been encouraged by the decision of the United Kingdom government to award WLL licenses at 10 GHz, with up to 200 MHz available per operator, for the purpose of providing WLL services to businesses. Typically, businesses currently subscribe to telephony services in multiples of 2 Mbps (or E1 lines, as they are often known); any credible WLL alternative would need to be able to provide one or more E1 connections per business.

Very little is known about such broadband systems, with the exception that Ericsson is developing a system in conjunction with Bosch called AirLine, which is being tested by Cable and Wireless Communications (previously Mercury), one of the 10-GHz license holders in the United Kingdom.

In principle, there is little difference between broadband and narrowband, other than a wider bandwidth signal is transmitted. The same access techniques, error-correction techniques, modulation techniques, and so on, all can be used. CDMA systems can become a little more difficult because once a high bandwidth (e.g., 2 Mbps) is multiplied by a typical spreading factor (e.g., 128), the resulting bandwidth of 256 Mbps requires baseband circuitry operating at this rate or higher to perform the decorrelation function. However, smaller spreading factors can be used, and baseband circuitry operating at above 100 MHz is now well established.

The Ericsson system is said to be capable of providing bandwidth on demand, which is basically a marketing term since all systems provide bandwidth on demand. What is meant is that a single user who at one moment might be using a 2-Mbps channel, at another moment, possibly during the same call, might use a 4-Mbps channel and then later perhaps only a 500-Kbps channel. That provides the user with flexibility while minimizing the use of the radio spectrum. A number of narrowband systems can provide such a capability. For example, DECT allows the user to increase or decrease the number of slots used in the TDMA structure, allowing a flexible bandwidth between 32 and 552 Kbps to be provided in multiples of 32 Kbps. The Lucent AirLoop system provides basic CDMA bearers with a capacity of 16 Kbps and allows users to multiplex simultaneously up to nine bearers (more in the future). It appears that the Ericsson broadband system will be CDMA based and probably will use an approach similar to the Lucent system (although with higher-capacity bearers) to provide bandwidth flexibility.

12.4 Summary of proprietary technologies

Table 12.6 summarizes the key characteristics of the different technologies considered viable for WLL provision in this chapter, excluding those

Table 12.6

Comparison of Proprietary Technologies

Services	Companies			
	Nortel	Tadiran	DSC	Lucent
Telephony	Yes, good voice quality	Yes, good voice quality	Yes, good voice quality	Yes, good quality
ISDN	For future release	For future release	Yes	Yes
Fax	Yes	Yes	Yes	Yes
Data	Yes, up to 64 Kbps	Yes, up to 32 Kbps	Yes, up to 144 Kbps	Yes, up to 128 Kbps
Videophone	No	No	No	No
Supplementary services	Wide range	Wide range	Wide range	Wide range
Multiple lines	2 lines	1 to 4 lines	2 lines	2- or 8-line units
Performance				
Range (radius)	15 km	6 km	5 km	4 km
Cells per 100 km^2	0.14	0.88	1.3	2
Capacity per cell, 2×1 MHz	10	8	8.5	5.75

technologies about which too little is known to date to provide a coherent analysis.

As can be seen, all four technologies provide a good range of basic services, including good-quality voice, fax, and data. The differences occur in the support of ISDN and high-speed data, with Lucent being particularly good.

In terms of performance, the Nortel equipment has substantially higher range than the equipment from the other manufacturers; if rural coverage is important, it probably will be preferred. DSC and Tadiran provide the highest capacity and so should be preferred in spectrum-constrained operations. The Lucent equipment is particularly poor in terms of capacity and should be adopted only if sufficient radio spectrum is available. That poor performance is somewhat unexpected given its use of CDMA, and it remains to be seen whether better performance actually can be achieved.

Figure 12.2 shows the sales achieved to date. As can be seen, the Nortel system has been the most successful. This is likely because it was one of the first systems to the market, and is not necessarily an indication of its superiority.

Chapter 13 compares the technologies from each of the different categories covered Chapters 10 to 12 and discusses the means whereby the most appropriate technology can be selected.

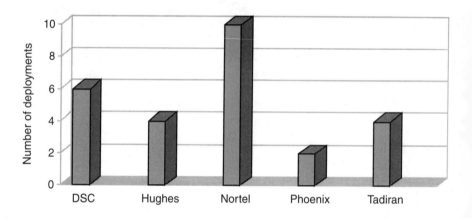

Figure 12.2   Sales of proprietary equipment at of the start of 1997.

13

Choosing the Right Technology

B Y THIS STAGE, you have been introduced to a wide range of different and confusing systems. This chapter helps guide the operator through choosing the most appropriate technology. To do that, the chapter is split into capacity, cost, and functionality comparisons.

Most operators will not be able to select from the full range of technologies. Spectrum availability will immediately limit the technologies that can be adopted, making the choice considerably simpler. Other factors will have a heavy bearing on the capacity required, the maximum cost, and the services that can be offered. Once all those factors have been taken into account, the choice of technologies is likely to be very small indeed.

13.1 Capacity comparison

Table 13.1 lists the number of voice channels that can be provided in a cell for each of the different systems.

Table 13.1

A Comparison of System Capacities

Technology	Voice Channels/ 2×1MHz/Cell	Comments
DECT	5.2	Can use adjacent channel in same cell
PHS	8	
CT-2	7	
TACS	3.3	
GSM	10	
IS-95	12	Sectorization can provide further gains
Nortel	10	
Tadiran	8	
DSC	8.5	
Lucent	5.75	

The information in Table 13.1 needs to be treated with caution. Some technologies, such as DECT, provide relatively few channels per cell but have inexpensive and easy-to-deploy base stations that can be used with a short range. By using more base stations, a DECT network potentially could provide higher capacity than a system such as DSC or Tadiran at an equivalent cost.

In general, the highest capacities per cell are provided by the proprietary technologies and the lowest by the cordless technologies. That is a deliberate design decision, because cordless is intended to be inexpensive to deploy, hence its simple and low cost but low-capacity base stations.

13.2 Cost comparison

Comparing costs requires an understanding of the range and the capacity of the different systems. The number of base stations for coverage and for capacity can be calculated as described in more detail in Chapter 17. Table 13.2 compares the relative ranges of the different systems.

Table 13.2 shows that only the cellular systems provide very high range. In rural areas, where the subscriber density is low, cellular systems

Table 13.2
A Comparison of System Ranges

Technology	Range (km)
DECT	5
PHS	5
CT-2	5
TACS	35
GSM	30
IS-95	30
Nortel	15
Tadiran	6
DSC	5
Lucent	4

are likely to be the systems providing the best economies. All the other technologies tend to provide similar ranges. That is because the cordless technologies operate at low frequencies, where propagation is better, but have low power, whereas the converse is true of the proprietary technologies. Hence, capacity, rather than range, will be a key differentiating feature between the different technologies.

In the calculation of cost, a key parameter is the number of base stations required. As will be shown in Chapter 17, the number of base stations tends to drive the sizing of all the other elements in the system. Given that, with the exception of the cellular technologies, the technologies have similar ranges, and the number of base stations will tend to be driven by whether additional cells are required, over and above those required for coverage, to provide adequate capacity.

A simple, first-pass estimate of the relative costs can be obtained by multiplying the cost per base station by the number of base stations required for each technology. Here is where the difficulties appear. The cost per base station is not publicized, is generally subject to confidentiality agreements, and in any case liable to significant change. For those reasons, it is not possible to expand the analysis on relative costs.

Some simple guidelines can be established if it is assumed that the cordless technologies all cost approximately the same and that the cost is approximately half that of the other technologies. The overall effect would be to make the cordless technologies the least expensive where they provided sufficient capacity without requiring additional base stations above those needed for coverage. Where additional base stations are required, cordless technologies may be slightly less expensive than the proprietary technologies, but the differences are small. PHS probably provides the best combination of low cost and relatively high capacity, with DECT not performing so well.

13.3 Functionality comparison

Table 13.3 compares the functionalities provided by the different systems. The table shows that only DECT and the proprietary technologies could be classed as providing a full range of services. Of the others, the cellular systems generally should be avoided because of inferior voice quality, while PHS and CT-2 can provide only limited data rates. Of particular importance in developed markets is the provision of multiple lines, which only Nortel, DSC, Tadiran, and Lucent are able to provide.

13.4 Summary

Selecting the most appropriate technology is a process of taking the constraints and using them as filters to derive a short list of technologies. The key filters are the following:

- The available frequency bands;
- The subscriber density;
- The competitive situation.

13.4.1 The available frequency bands

Table 13.4 shows which technologies can be adopted, depending on the frequency band available.

Table 13.3
Comparison of System Functionalities

Technology	Voice quality	ISDN	Data	Services	Multiple Lines
DECT	Good	Yes	552 Kbps	Good	Up to 12
PHS	Good	No	32 Kbps	Limited	No
CT-2	Good	No	32 Kbps	Limited	No
TACS	Poor	No	No	Limited	No
GSM	Medium	No	9.6 Kbps	Good	No
IS-95	Medium	No	9.6 Kbps	Good	No
Nortel	Good	Future release	64 Kbps	Good	Yes
Tadiran	Good	Future release	32 Kbps	Good	Yes
DSC	Good	Yes	144 Kbps	Good	Yes
Lucent	Good	Yes	128 Kbps	Good	Yes

Table 13.4
Equipment Available in the Different Frequency Bands

Frequency Band	Technology
800/900 MHz	Analog cellular, GSM, IS-95, CT-2
1.5 GHz	Tadiran and some other proprietary manufacturers
1.7–2 GHz	DECT, PHS, GSM, and IS-95 variants
2–2.5 GHz	DSC, Tadiran
3.4–3.6 GHz	Nortel, Tadiran, Lucent
10 GHz	Emerging technologies
>10 GHz	MVDS technologies

13.4.2 The subscriber density

Table 13.5 shows which technologies are most appropriate for which densities.

13.4.3 The competitive situation

Table 13.6 shows which technologies are most appropriate for which competitive situations.

So an operator with a license in the 3.4 to 3.6 GHz band in a suburban environment in a first world fully liberalized competitive situation would

Table 13.5

Equipment Suitable for Differing Environments

Density	Technology
Extremely low-density rural	Analog cellular, GSM, IS-95
Rural	Analog cellular, GSM, IS-95, all proprietary technologies
Suburban	All technologies
Urban	DECT, PHS, CT-2, and all proprietary technologies
Metropolitan	DECT, PHS, CT-2, and all proprietary technologies

Table 13.6

Equipment Suitable for Differing Competitive Situations

Density	Technology
No telecoms provision	Any
Eastern Europe environment	All cordless and all proprietary technologies
First world limited competition	All cordless and all proprietary technologies
First world fully liberalized	DECT, Nortel, DSC, Tadiran,and Lucent

start with Nortel, Tadiran, Lucent, and DSC, according to the frequency band. All those would be suitable for suburban areas, and for the competitive environment. It would then be a case of trading off the better data capabilities of some systems against the other systems.

Alternatively, an operator with a license in the 1.7 to 2 GHz band operating in a mostly rural environment where the competitive situation is best characterized as Eastern European would start with the possible technologies as DECT, PHS, GSM, and IS-95 based on the frequency band. The operator then would narrow that down to GSM and IS-95, according to the need for high range but would discover that neither met the requirements for the competitive situation. In that case, a quality IS-95 installation with high signal strength might provide satisfactory voice quality.

Part V

Deploying Wireless Local Loop Systems

P ART V DEALS with the practical deployment of WLL systems, from applying for a license to run a system through designing the service, developing the business case, rolling out the network, and providing on-going support.

WLL networks are based around cells that deliver signals to receivers and as such are broadly similar to cellular systems. Nearly all the aspects of deploying the system are closely related to those for cellular. Readers also may want to draw on other references and their own experience in the cellular industry. Further, each country and each network is different; the information provided here can be only a guide.

14

Getting a License

THE FIRST STAGE in gaining a license is noticing that there is an opportunity to gain one. Chapter 4 introduced details about potential markets in a number of different regions. Each operator will have its own view as to which types of markets it favors, often based on experiences.

14.1 The opportunities

The simplest opportunities to spot are those in the home country. Operators currently providing a cellular or a trunked network are well placed to notice whether there appears to be a substantial unmet demand for WLL. Individuals in the telecommunications industry may be well placed to start their own companies to take advantage of opportunities.

Opportunities in foreign countries are more difficult. Most large telecommunications investors will hire individuals (so-called license spotters) in a range of countries to obtain information as to whether opportunities are likely in the country.

One simple way to spot opportunities is to monitor the actions of the government department administering licenses for radio or telecommunications applications. Often, forthcoming competitions for licenses are announced publicly, allowing an operator to investigate a potential opportunity.

Operators that want to follow a formalized approach might consider the procedure described in the following sections; where countries are segmented into the following categories: developing countries, Eastern European countries, and first world countries. In effect, the procedures described below form a sieve into which pass all countries and out of which fall the most attractive countries. Each category is examined in detail to discover whether additional factors affect its attractiveness.

14.1.1 Developing countries

Information should be gathered on the following factors:

- The current teledensity;

- The GDP per head;

- The political stability of the country;

- The telecommunications environment in the country, including the number of operators and the relative competition among them;

- The presence of international aid funding;

- Projected growth figures for the country.

Of key interest are countries with a relatively low teledensity compared to the GDP per head. Such countries are prime candidates for WLL, as long as there are no concerns over the stability of the country or the competitive environment. Information on GDP per head can be obtained from a range of sources, for example, *The Economist*; information on teledensity is available from organizations such as the ITU. Some typical figures for South America are shown in Table 14.1.

Table 14.1

GDP and Teledensity Figures for South America

Country	GDP/head (U.S. $)	Teledensity (lines per 100)	Teledensity/ (GDP/head) × 1,000
Guatemala	1,100	10.0	9.1
Colombia	1,472	10.0	6.8
Dominican Republic	1,143	7.4	6.5
Costa Rica	2,667	13.0	4.9
Panama	2,400	11.5	4.8
Bolivia	625	2.9	4.7
Uruguay	4,333	18.0	4.2
Venezuela	2,810	11.0	3.9
Chile	3,143	11.0	3.5
El Salvador	1,333	4.2	3.2
Paraguay	1,400	3.1	2.2
Argentina	7,529	14.0	1.9
Peru	1,957	3.3	1.7

In Table 14.1, the second and third columns are relatively self-explanatory. The numbers in the fourth column are somewhat unusual and are obtained by dividing the teledensity by the GDP per head. The result is multiplied by 1,000 to make the numbers more legible. Column 4 is a measure of whether a country is underserved or overserved relative to its GDP. For example, a country with better provision of telecommunications than would be expected for its GDP would score highly in the fourth column. An operator would be most interested in countries that are poorly served relative to their GDP, because they would have a high demand for telecommunications coupled with the capability to pay for the services. Table 14.1 is ordered by the column 4, so the countries at the bottom of the table are the most interesting.

For example, Peru has a GDP per head of around $2,000 but only 3.3 lines per 100 people. That compares badly with Guatemala, which has almost half the GDP per head but 10 lines per 100 people. Countries like Peru, Argentina, and Paraguay should be investigated in more detail to determine why the penetration compared to GDP is so low.

14.1.2 Eastern European countries

Information should be gathered on the following factors:

- The current teledensity;

- The average delay before a line is installed;

- The GDP per head;

- The projected growth;

- The number of competitors;

- Government demands for local shareholdings and manufacture;

- A high penetration of cellular compared to GDP per head.

The key here is to look for countries with either a low teledensity coupled with a high GDP or a high delay for line installation coupled with a high GDP. A high penetration of cellular relative to the GDP also could be an indication of substantial unmet demand for fixed lines. In the Eastern European countries, competition is likely to be a significantly more serious issue than it is in the developing countries. The Eastern European countries also are often keen to use new networks as an opportunity to increase the skills in their country and often insist on local shareholdings and manufacture. Such insistences should be scanned carefully to see whether they are problematic. Some data on GDP and waiting lists for Eastern Europe is provided in Tables 14.2 and 14.3.

Table 14.2 provides data presented in the same manner as that for the developing countries. It is clear that telephone provision is relatively good, scoring a minimum of 4.5 in column 4, compared to 1.7 for the developing countries. Another way of looking at the problem is provided in Table 14.3, where the number of people waiting per 100 is presented (column 3) and normalized against GDP (column 4).

Waiting list figures typically are less useful. For example, Greece looks attractive on the waiting list because the GDP per head is high compared to other countries; thus, the relatively small waiting list results

Table 14.2
GDP and Teledensity for Eastern Europe

Country	GDP/head (U.S. $)	Teledensity per 100	Teledensity/ (GDP/head) × 1,000
Romania	1,087	12.3	11.3
Turkey	1,984	20.1	10.1
Russia	2,223	16.2	7.3
Ukraine	2,137	15.0	7.0
Czech Republic	3,200	20.9	6.5
Greece	7,400	47.8	6.5
Poland	2,158	13.1	6.1
Belarus	3,100	17.6	5.7
Hungary	3,800	17.0	4.5

Table 14.3
GDP and Waiting Lists for Eastern Europe

Country	GDP/ head (U.S. $)	Number waiting per 100 people	Waiting list/ (GDP/head) × 1,000
Romania	1,087	5.7	5.2
Russia	2,223	7.3	3.3
Ukraine	2,137	6.9	3.2
Poland	2,158	6.3	2.9
Belarus	3,100	7.0	2.3
Czech Republic	3,200	6.0	1.9
Hungary	3,800	7.0	1.8
Turkey	1,984	1.1	0.6
Greece	7,400	2.0	0.3

in a low overall score, which should indicate a particularly underserved market. However, the tables do agree on many countries. For example, both Tables 14.2 and 14.3 agree that Romania should be avoided because the GDP is so low that additional telecommunications probably is not affordable by the population there.

14.1.3 First world countries

Information to obtain includes the following factors:

- Annual price reduction of services offered by the PTO;
- Number of competitors;
- Annual spending on telecommunications.

Essentially, the key is to look for uncompetitive PTOs, where an efficient WLL operator could gain market share rapidly. Where the PTO has become competitive, that is normally marked by falling prices. Falling prices also reduce the call charges that the WLL operator can offer. The most attractive markets are those where prices have been stable or even have risen. The competitive environment is critical, and the total spending allows the total market to be sized.

14.2 The license application

To be awarded a license, it is normal to submit a license application. The scope and the content of a license application vary dramatically, depending on the situation. For example:

- Where the license award is by auction based on the highest price paid, the license application typically need only provide supporting proof of registration of the operator, possibly including financial statements depending on requirements.
- Where the license application is noncompetitive, the document needs to explain clearly why the operator should be given the license.
- Where the license award is by a process known as a "beauty contest," a substantial and well-written document is required. In a beauty contest, the government typically sets a number of criteria against which the bids are evaluated, such as improvements in

telecommunications in the country, provision of jobs to the country, rollout plans, and efficiency of the use of the radio spectrum. Each criterion needs to be addressed fully to convince the regulator that the operator should be awarded a license.

The following general factors should be included in the license application:

- Introduction;

- Overview of the operator, including ownership and turnover;

- Details of the service proposed;

- A technical description of how the network will be realized;

- A demonstration that the radio spectrum required is being used efficiently;

- Simplified business plans showing the justification for building the network;

- A network build plan showing the speed with which the network will be built;

- A summary of the advantages to the country;

- Appendices, including coverage maps, detailed technical information (where appropriate), details of the frequencies required, and interference assessments with other users of the radio spectrum (where required).

The most important fact to understand is what the regulator or the government wants to get from the license award. Possible government objectives include the following:

- Increasing the teledensity at a higher rate than the PTO could achieve;

- Providing telephony to rural regions;

- Stimulating local industry and improving local skills;

- Gaining a cash injection through the sale of radio spectrum;

- Providing competition to the PTO, thus driving down prices;
- Providing higher bandwidth services, thus allowing further economic development.

The criteria at the start of the preceding list tend to be most prominent in the developing and Eastern European countries; those criteria at the end more likely in the first world countries. Clearly, understanding the objective and tailoring the license application and business plan in a manner that focuses directly on maximizing the benefit received by the government will go a long way toward increasing the possibility of the license application being successful.

On a more prosaic level, it should be remembered that one of the key concerns of the government officials who decide whether an application should be granted will be not to make an inappropriate decision. They will be looking for reassurance that the organization is competent and will perform as specified. Presenting a substantial and well-formatted application document will go some way to reassuring them that the application is genuine and the organization competent.

14.3 Radio spectrum issues

14.3.1 Radio spectrum management

All WLL systems require radio spectrum. Without radio spectrum, there can be no transmissions. Radio spectrum, however, is a scarce resource with numerous competing demands, managed by government-appointed bodies. Many potential WLL networks fail because of an inability to gain sufficient spectrum. This section is designed to help operators understand better the spectrum management process and hence be better able to deal successfully with the spectrum manager. This section is quite long: a lot is happening in spectrum management and the changes will be of critical importance to prospective operators. The section first explains the key players in spectrum management, then looks at why spectrum management is changing and the likely results. It ends with a discussion as to how spectrum management should be performed and the implications to the WLL operator.

The world of radio spectrum management comprises a number of important institutions. The most important is the ITU (previously known as the CCITT, from the French for *central committee for telephony and telegraphy*). The *radio division* (ITU-R) is responsible for drawing up guidelines for international spectrum use. The ITU has no formal power, but its importance is widely recognized and nearly all countries abide by its recommendations. The work of the ITU is conducted through biannual *world radio conferences* (WRCs). At those conferences, proposals are discussed for changes to the manner in which spectrum is used. The meetings often are fraught with tension as each country tries to negotiate a settlement that meets its own interests. At the end of each conference, the ITU recommendations are updated. An ITU recommendation looks something like the stylized version shown in Table 14.4.

Each recommendation forms one page of a large book covering the entire radio spectrum band from 10 kHz through to around 300 GHz. For ITU recommendations, the world is divided into three regions:

- Europe, Russia, and Africa;
- North America and South America;
- The rest of the world.

For each region, a primary allocation is shown in uppercase letters. The primary allocation has key rights over the spectrum, can use the spectrum without concern for the interference it generates (within the frequency band designated), and can expect not to suffer interference from others. There also may be a secondary allocation, which is shown

Table 14.4
A Stylized Example of an ITU Recommendation

Frequency Band 1.455 GHz–1.460 GHz		
Region 1	Region 2	Region 3
FIXED	FIXED	SATELLITE
aeronautic (note 234)	MOBILE	fixed

in lowercase letters. The secondary allocation has to coexist with the primary allocation, must not interfere with the primary allocation, and must expect to suffer interference. Secondary allocation is known as sharing. Sharing often works well between, for example, fixed links and satellite links, because both are stationary and can be planned around each other. In some cases, there are two primary allocations, which must interwork on a mutual status.

It is worth noting that the ITU-R recommendations actually are somewhat vague. For example, the Region 2 allocation shown in Table 14.4 allows almost any use of the radio spectrum, from private mobile radio through microwave links, wireless local loop, and so on.

As regards WLL, the ITU is important in providing guidance for bands that can be used worldwide, allowing economies of scale to be achieved. However, there is no ITU-R category for WLL, so countries will be seeking a FIXED status in a number of key bands, such as 3.4 to 3.6 GHz. Whether that will be achieved remains to be seen.

Below the ITU may be another layer of spectrum management, for example, the CEPT represents Europe and Eastern Europe, and Africa has an embryonic spectrum management body. Bodies like the CEPT have as an objective the standardization of spectrum use throughout Europe, easing cross-border issues, and providing a strong voice in WRC meetings. CEPT recommendations are much more specific than ITU recommendations and might include a specific allocation for WLL. Like the ITU, the CEPT has no official power, but its decisions are respected by its members, who realize that doing so is in their best interests.

CEPT decisions are reached by consensus. Typically, one of the member countries or a standardization body provides an input document to CEPT stating an interest in a pan-European allocation for a particular application. Study teams then are appointed within CEPT to look into the issue. The teams assess the total spectrum requirements for the new service and the most appropriate frequency bands where the service might be located. They then approach each of the members to understand whether the members might be able to make radio spectrum available for the application. Finally, they reach a recommendation as to what spectrum should be allocated. All the members then vote on the recommendations. In some cases, members might choose not to accept the recommendation because of existing use of the spectrum. Generally, if the proposal has widespread endorsement, CEPT accepts the allocation, although it may be noted that in some countries the allocation will not be available in the near future.

The CEPT also has been attempting to produce a common European radio spectrum band plan. That has proved to be a task fraught with difficulty. The existing uses of the bands are varied throughout the CEPT countries, and few members are prepared to change allocations (which would require users to replace their radio equipment) for the purposes of harmonization. The CEPT has discovered that harmonization is possible only in the lesser used parts of the spectrum (e.g., above 3 GHz) and then only if the harmonization is a long-term goal (in this case, harmonization is intended by 2008).

Below the multinational bodies lie the national spectrum managers, which are part of the government and administer both allocation and assignment on a national level. Their role includes making decisions as to the use for a particular band within ITU and CEPT guidelines, deciding which operator will have access to the band, monitoring the band usage, and resolving cases of interference. The national spectrum management bodies do have official power.

In general, a spectrum manager has three tasks:

- Fair distribution of spectrum where there are a number of applicants for the spectrum (assignment);

- Clearance of bands for new applications or the migration of users to more efficient equipment as it is developed (clearance);

- The decision as to how much spectrum to give to a particular use (allocation).

Currently, those decisions typically are made administratively and not by the use of any clear and objective measures. A spectrum manager will carefully consider the evidence, normally provided in a technical and economic form, and will make a judgment as to the most appropriate decision.

For example, assignment normally is conducted either on a first-come first-served basis, in which the spectrum is given to applicants until there is none left or by a beauty contest, in which applicants are judged according to criteria such as their financial standings and previous track records. Clearance occurs in two stages. First, after a long process of consultation, a spectrum manager makes a judgment that clearance is necessary. Then the incumbent is given a long period, typically 10 years or more, to move, which allows their equipment to depreciate fully. Allocation is performed mostly on a first-come first-served basis, with the advantages of the contending uses weighed in an administrative manner.

Spectrum management is an issue only when spectrum is congested (otherwise, there is sufficient spectrum to meet all apparent needs) and when use of spectrum can be improved through greater efficiency.

14.3.1.1 Congestion

Most users of radio spectrum have some intuitive understanding as to when congestion is occurring. Rules of thumb adopted by the users include when no channels are available at the site on which they apply for a license and when license rejection rates reach a significant level as perceived by the user. Although these rules of thumb undoubtedly are useful, to be used consistently, congestion should be defined on a more rigorous basis. A number of possible definitions have been considered, of which the most promising is the following:

A band is congested when there is an excess of demand over supply. The level of congestion is the cost to the potential users of using alternative communications means where necessary.

That measure encapsulates the level of congestion and the importance of that congestion to the country. Where there is no congestion, the cost is zero. The difficulty with such a measure is that the excess demand often is not clearly visible: some users do not apply for frequencies because they are sure that their request will not be granted. Quantifying the additional costs faced by users also can be difficult because of commercial confidentiality. Despite those difficulties, approximations typically can be used in deriving this measure of congestion.

14.3.1.2 Efficiency

To understand how the term *efficient use of the spectrum* should be defined, it is necessary to consider the objectives of the spectrum manager. The spectrum manager manages a valuable and finite resource on behalf of the government and hence the country. Good spectrum management can be defined as maximizing the value that the country achieves from this scare resource, leading to this definition:

> Spectrum is used efficiently when it is used in such a way that the economic value to the country of its use is maximized.

That definition is somewhat unsatisfactory, because it is extremely difficult to determine whether the economic value indeed has been maximized. Nevertheless, some basic principles that can be used to help judge efficiency can be established. The economic value achieved from a band of spectrum is increased, approximately proportionally, as the number of users or the traffic generated per user is increased (as long as the quality of service is maintained). The economic value is decreased as the cost of using the spectrum rises, for example, as equipment prices become more expensive. Those principles show that, as an approximation, technical efficiency can be used to help understand whether efficiency could be enhanced. There are a number of possible measures of technical efficiency, but, for the purposes of spectrum management, one study [1] suggests that the most appropriate measure is as follows:

> Technical efficiency is a measure of the information transmitted per megahertz per day, nationwide by a particular use or user.

As an example of this, approximations suggest that *private mobile radio* (PMR)[1] users transmit only one-quarter of the information transmitted per megahertz by cellular users. Hence, were PMR users accommodated on cellular, only one-quarter of the spectrum currently used for PMR would be required. To understand whether that would improve effi-

---

1. Known as *specialized mobile radio* (SMR) in the United States.

ciency, it is necessary to look further at the value gained by PMR users by examining the functionality they achieve from their system. This topic is discussed in detail in [1], but suffice it to say here that for some users with demanding requirements, efficiency is greater if they remain with PMR, despite a lower technical efficiency; for most users, however, a higher efficiency of use would be expected on cellular.

There is a further dimension to efficiency. The spectrum is used most efficiently when both allocation and assignment are performed in such a manner as to divide the spectrum among the users who generate the greatest economic value to the country. Typically, technical efficiency seeks only to improve the use of spectrum within the existing allocation framework.

14.3.1.3 New Trends

The emergence of a number of new trends and pressures in the use and regulation of radio spectrum is making the task of spectrum management increasingly difficult. These of those trends are discussed here: first, those originating from the users of radio spectrum and, second, those originating from the government. As regards the users, there are the following problems:

- The continuing growth in demand among existing spectrum users;

- The dramatic increase in new applications requiring radio spectrum;

- The increasing profit that can be made from possession of a license to operate a telecommunications network in certain frequency bands, resulting in increasing pressure on the spectrum manager to make such licenses available.

As regards the government, a number of trends are taking place, such as the need for revenue, fairness, and increased competition. Those overall objectives, which affect all parts of government policy, also have an impact on spectrum management policy, affecting the manner in which the spectrum manager works. Important trends include the following:

- The need for greater openness in government;

- The trend toward deregulation of the telecommunications sector and the increased competition that has resulted;

- The use of spectrum as a fund-raising resource by some governments;

- The possibility of privatization of the spectrum manager.

14.3.1.4 Growth in demand

One major problem facing spectrum managers is the ever increasing growth in demand. For example, in many countries, the growth in cellular radio subscribers is some 60% per year. Even in the world of PMR, whose demise has been predicted often, annual growth continues to exceed 6% in the United Kingdom. To some extent, that growth in demand can be met by advances in technology.[2] However, at best the advances in technology cancel the growth of users, and in the high-growth world of cellular they do not come close to accommodating growth. The growth in demand can only increase the pressure on the spectrum manager to find additional spectrum.

14.3.1.5 Dramatic increase in new applications

As well as growth in existing applications, recent years have seen a dramatic increase in the overall number of applications vying for spectrum. That has been caused mainly by improvements in technology, which have made certain applications possible, and by reducing equipment cost, which have made existing applications economically viable in new deployments. New applications that have arisen recently include mobile-to-satellite communication systems, such as Iridium and Globalstar, and, of course, WLL systems.

---

2. For example, PMR channel bandwidth has tended to be halved approximately every 10 to 15 years (25-kHz channels in the 1970s, 12.5-kHz channels in the 1980s, and 6.25-kHz channels in the 1990s). Each halving approximately doubles capacity. An annual growth rate of 6% requires that capacity be doubled every 12 years to accommodate growth. Thus, in this case, technical advances are meeting the demand for growth. That is not true in cellular, where growth of 60% per year outstrips technical advances, which achieved a capacity increase of around 100% in the last 10 years.

14.3.1.6 The profits to be made from spectrum use

Organizations offering services based on the spectrum they have been assigned are generating increasingly large profits. For example, cellular operators throughout Europe are seeing their values increase rapidly. The realization that access to radio spectrum is becoming increasingly profitable is causing more commercial operators to bring increasing pressure to bear on spectrum management organizations for access to the appropriate spectrum, both through direct lobbying to the spectrum manager and indirect lobbying through elective bodies. The pressures brought to bear in that manner typically are much more forcefully expressed than those where profit is not the key driving element.

14.3.1.7 The need for openness

Government departments, especially in the developed world, are coming under increasing pressure to act in an open manner, with key decision documents made publicly available and the evaluation criteria for major decisions publicized. That can pose problems for spectrum managers using the current judgment-based methods for allocating and assigning spectrum, because it often is difficult to clearly and simply justify a decision, often reached after a long process of weighing contradictory claims. Using an approach such as an auction, where the winner is universally understood to be the entity paying the most, provides a much more open and transparent means to assign spectrum.

14.3.1.8 Deregulation and increased competition

In many developed countries there is a trend to deregulate the telecommunications sector. Such moves are attractive to the state and to the population because increased competition often brings about better service and lower prices. However, deregulation can cause problems to the spectrum manager. Increased competition typically brings with it requirements for more spectrum without actually offering any additional services to the public. Additional spectrum must be provided to the new operators on a fair and equitable basis compared with the assignment to the incumbent.

14.3.1.9 Use of spectrum as a government revenue-raising resource

It has been known for many years that radio spectrum is a valuable resource, but recently governments have discovered that they can capture part of the value through selling the radio spectrum in some form. In an era when governments are under pressure to balance budgets without increasing taxes or reducing social provision, pricing radio spectrum, along with other revenue-raising techniques such as privatization, is seen as an important supplementary source of income. That can increase the pressure on spectrum managers to move toward revenue-raising economic spectrum management approaches.

14.3.1.10 Privatization of the spectrum manager

In some developed countries, there is a desire to reduce the overall size of government by returning as many functions as possible to the private sector. That normally is achieved through the process of privatization. Such a move could change the way spectrum managers operate.

14.3.2 Modern allocation and assignment methods

A number of economic techniques could be used by spectrum managers. They are listed next, along with details of which of the trends identified in Subsection 14.3.1 they are able to address. Each technique then is discussed in detail.

- Economic value of spectrum analysis estimates the value that the country achieves from each different use of the spectrum and hence gives some signals as to which uses should be preferred. This technique helps overcome the difficulty of the increasing numbers of new applications.

- Administrative pricing increases the annual license fee required for the spectrum to a level that encourages users to make careful choices about the efficiency with which they use the spectrum. This technique helps overcome the growth in demand and the pressures on assignment, as well as all the difficulties associated with government policies.

- Trading allows users a range of rights to sell or lease spectrum and possibly to subdivide their assignment. This technique helps overcome the growth in demand and new applications, helps overcome the profit-driven pressure on assignment, provides greater openness and allows more deregulation of the telecommunications sector.

- Competitions such as auctions determine to whom large blocks of spectrum should be assigned. This technique overcomes the growth in demand, the profit-driven pressure, and all the government objectives.

Table 14.5 shows which tool can be used to solve which general task. In some cases, it is clear that certain tools can or cannot achieve particular objectives. In others, it depends very much on the objective. For example, administrative prices can be used to judge the value of the spectrum and influence allocation decisions in some situations. In many cases, however, overriding issues will prevent this from occurring effectively.

Different tools can be used in different parts of the radio spectrum, and a number of tools can be used in conjunction. The detailed aspects associated with the application of each of the tools are discussed next.

14.3.2.1 Economic value analysis

Economic value analysis estimates the total value generated in the economy of a country as a result of activities associated with radio spectrum. For example, profit generated by cellular operators, the salaries paid to their staff, the equipment they purchase, where equipment is manufactured in the country, and the salaries paid to sales staff in retail outlets all would be counted. That allows the value added by each different use of

Table 14.5
Application of Economic Tools

	Allocation	Assignment	Clearance
Economic value analysis	Yes	No	Yes
Administrative pricing	?	Yes	Yes
Trading	?	Yes	?
Competitions	No	Yes	?

the radio spectrum to be compared, so it can be determined which applications add the most value to the country.

Advantages By understanding the economic value, more spectrum can be allocated to those applications generating high levels of value. Thus, the allocation process becomes better informed. Interestingly, one of the main benefits of the work in the United Kingdom is that it has indicated the value of spectrum to government ministers, which has helped in the consideration of subsequent legislation designed to improve its use.

Disadvantages The main problem with economic value analysis is the difficulty in performing the work to any reasonable level of accuracy. A significant amount of data must be collected, for example, the total number of PMR users, the number in each category of use (e.g., taxi drivers), the cost of the PMR equipment in use, and the cost of substitute equipment. Experience in the United Kingdom has shown that typically it is possible to gather much of the required data but that the collection can take some time. With adequate data, the economic calculations are relatively straightforward.

14.3.2.2 Administrative pricing

Most administrations set license fees at a level that covers the cost of administering the spectrum. Under administrative pricing, that link is broken and prices are set at a level that approximates the market value of the spectrum. Prices ideally are set at a level where supply and demand match, and hence congestion is eliminated. At those levels, users then would have incentives to release unused or underused spectrum, to consider alternative services or uncongested frequency bands, and to implement more spectrally efficient technologies. The prices would not necessarily be relatively static, like the current license fees, but would change to reflect both changes in demand (e.g., the increase due to a new application being provided over radio) and changes in technology (e.g., new narrowband radio systems). The spectrum manager would need to carefully monitor supply and demand conditions to be able to change the price.

There is one major work that describes the process of setting administrative prices [2]. Detailed economic arguments explain how prices are set, but fundamentally a user in a congested band where demand exceeds supply needs to see that investment in an alternative that uses less spectrum would lower costs. That currently is not the case, so the role of pricing is to change the costs of the user such that the alternative becomes attractive.

More explicitly, once pricing is introduced users will be faced with the cost of the current equipment or service (as they are at present) plus a new cost that is the annual administrative price over the equipment lifetime. The alternative is for users to move to a different form of equipment or service that uses less spectrum. Because it uses less spectrum, the administrative cost component is lower than that of the current equipment. Even if the alternative is more expensive before pricing is introduced, it becomes less expensive after pricing.

For example, a PMR user might purchase narrowband equipment, move to a shared PMR system, or move to cellular systems. The cost of each alternative can be calculated. If, say, narrowband equipment is the lowest cost, then the administrative prices should be such that the cost of the new equipment is offset by the savings in administrative prices over the equipment lifetime.

Advantages The key advantage of administrative pricing is that it can encourage existing users to think carefully about the amount of spectrum they require, preventing hoarding and providing an incentive to use narrowband equipment. A secondary advantage is the raising of additional resources for either the radio spectrum management or government use.

Disadvantages Calculating prices is difficult and unlikely to be accurate. The government needs to monitor carefully the effect of the prices and adjust them accordingly, which requires substantial skill.

14.3.2.3 Trading

Spectrum trading is a means to develop a real market in spectrum, as opposed to the pseudo-market created by administrative pricing. By allowing trading between users of spectrum, prices can fluctuate rapidly in accordance with the actual supply and demand, as opposed to that perceived by the administrator. Many of the goals are the same as for administrative pricing in that because the spectrum has a value users are inclined to use it efficiently. The incentive is not the stick of annual fees but the carrot of potentially large cash gains through sale.

There are many options with trading, including the simple sale of a license to another entity, the ability to divide a license and sell parts of it to another entity, and the ability to lease the license to another entity. One or all of these options could be provided. The introduction of trading often requires major new legislation and the design of an extended transitional program as tradability is extended across various areas of the spectrum. Important transitional steps typically include the following:

- Setting up a publicly accessible spectrum register;

- Introducing spectrum trading rights gradually and differentially across different areas of spectrum;

- Limiting capital gains that might be earned by license holders, possibly including auctions and increases in charges;

- Introducing compensation for repossession of spectrum.

Advantages Trading has a number of advantages over administrative pricing, for example:

- The true market value is realized for the spectrum.

- Trade in spectrum typically will be able to take place more rapidly than the administrator could reassign spectrum.

- Organizations purchasing others readily would be able to keep any spectrum involved.

Disadvantages Trading has a number of disadvantages, for example:

- Windfall gains may be made by those who happen to hold spectrum at the time that trading is introduced.

- The value of the spectrum may not be realized by some users, so the desired signals to improve the efficiency of spectrum use may not be seen.

- The state does not capture any revenue.

- The market may be difficult to administer and regulate, resulting in distortions of the market and problems when spectrum needs to be repossessed.

14.3.2.4 Competitions

Competitions can be used where there are large blocks of spectrum and more competing organizations that require the spectrum than there is spectrum available. The most straightforward form of competition is the price auction, whereby the bidder offering the highest price is awarded the spectrum. Other techniques, include beauty contests, in which qualitative judgment is used, and service auctions, in which the bidder offering the best service to the user is the winner. There are a number of auction designs, and the whole area of running an auction requires careful study. Much of this study has now been performed and documented for the *personal communications service* (PCS) auctions in the United States, which have shown that, if conducted carefully, competitions can be a successful tool for assigning spectrum. Where auctions are being used, most spectrum managers are now adopting the U.S. PCS auction methodology, thus avoiding the need to partake in auction design. Auctions are widely used for cellular licenses around the world but typically not for other applications.

Advantages   Some of the advantages of auctions are the following:

- They are transparent and visibly fair.

- They are cheap to operate compared with the alternative assignment methods, especially when auction designs proved by others are adopted.

- They raise money for government.

- They can be devised to take into account a variety of valid concerns other than just price paid.

- They give a market test of the value of spectrum, which may be of value for devising policy in areas where auctions are not appropriate.

Disadvantages Spectrum auctions are not suitable for small or low-value uses, such as local PMR and individual fixed links, because of the relatively high transaction costs in conducting auctions, problems in defining a product to sell, and the intermittent nature of demand. However, there might be scope for auctioning national PMR and fixed-link bands to organizations that would provide local spectrum licenses to individual users.

14.3.2.5 Spectrum management and WLL
For WLL, it is likely that some of these tools will be used. The simplest tool to use for WLL is competitions, typically in the form of auctions. That is because much of the spectrum used for WLL will be empty and there may be competing applicants. Auctions can help make a decision between different WLL operators.

Spectrum pricing also may be an issue. If the spectrum is provided at no cost, and spectrum pricing is used to control the growth in fixed-link demand, it will be hard for the spectrum manager to defend not applying spectrum pricing to WLL. Fortunately, the prices applying to fixed links are relatively low, so such pricing may not be problematic.

In some countries, it may be possible to avoid pricing altogether. Where the government is eager to introduce competition to the PTO, it may be prepared to waive spectrum fees to allow the new WLL operator to compete as effectively as possible, which was the case in the United Kingdom.

14.3.3 Implications of new spectrum management techniques for the WLL operator
The discussion here on spectrum management has shown that shortage of spectrum is still very much a problem. Predictions that digital modulation and microcells would lead to a world where spectrum scarcity was a thing of the past have not come about [3]. Digital modulation has offered only slight improvements in capacity over analog modulation, and microcells remain expensive and are deployed only where absolutely necessary. A shortage of spectrum leads the spectrum manager to consider carefully how it should be distributed and has important implications for the WLL operator.

For the operator, the implications of a spectrum shortage are as follows:

- Spectrum is available only at higher frequencies, preventing the use of many of the available technologies and restricting the range and coverage to LOS propagation.

- It is important to deploy the more technically efficient technologies, such as CDMA, to maximize capacity and demonstrate to the spectrum manager that the spectrum is being used as efficiently as possible.

- High-bandwidth applications, such as 384-Kbps bearers, will require significant amounts of spectrum, requiring a more dense infrastructure. Thus, although a technology may offer a high data rate bearer, lack of spectrum may prevent operators from deploying that service unless they are prepared to build a high density of cell sites.

For the spectrum manager, the implications of a shortage of spectrum are the following:

- Users must be required to demonstrate the economic value of their application to retain spectrum. Statistics kept by the operator showing the value added to the economy by their business and their success in meeting coverage and competition objectives will be valuable.

- In distributing the spectrum, market forces such as auctioning and pricing are likely to be used to ensure that the most economically efficient use of the spectrum is achieved.

Some spectrum managers may feel that it is their task to advise WLL operators on which technology they should adopt. To date, there has been little tendency to do that for WLL, but in some countries it is likely. WLL operators in those countries will need to determine whether the mandated technology is appropriate; if it is not, the operators will have to petition the government to be allowed to use a different technology.

Another potential area for obligation is coverage. Regulators can see coverage obligations as a way of preventing cream-skimming and ensuring that the operator remains serious in its investment in the network. As argued in Chapter 5, that is a form of USO requiring cross-subsidy and should be avoided in a competitive environment. Nevertheless, it seems extremely common. In practice, few spectrum managers will rescind a license when a network provides 50% coverage rather than the required 60% coverage, so there will be some latitude in not meeting that obligation. (In any case, the lack of real power to enforce such an obligation makes it inappropriate for the spectrum manager to make it part of the license condition.)

References

[1] NERA, Smith, *Review and Update of 1995 Economic Impact Study*, Radiocommunications Agency, London, U.K., 1997.

[2] *Study into the Use of Spectrum Pricing*, U.K. Radiocommunications Agency, April 1996. (Also available on the Internet at http://www.open.gov.uk/radiocom.)

[3] Calhoun, C., *Wireless Access and the Local Telephone Network*, Norwood, MA: Artech House, 1992.

15

Choosing a Service Offering

D IFFERENT TECHNOLOGIES have different capabilities, as was indicated in Part IV. It is important that the appropriate mix of capabilities, or services, is selected for the target market. If the total service package is insufficient, customers will migrate to competitors who offer a better service. If the service package is excessive, the system cost will be high and competitors can attract customers by offering a lower-cost service.

Choosing the right mix is a task for the marketing department. They should understand the services available, the services that the potential customers say they require, and the services that customers reasonably might be expected to require but do not yet realize they need, given global trends in telecommunications and local trends in growth and modernization. The marketing department needs to understand how much customers likely will pay for each of the services, the distribution of demand, and the strategy that the competition is likely to adopt. One of the key trends driving increased service is the need for Internet access.

221

Internet users, as explained in Chapter 2, often require an additional line coupled with relatively high-rate data capabilities. That will drive WLL systems to offer at least two lines in all but the least developed countries.

More details about how to determine which service offerings to adopt based on predictions of penetration and revenue are provided in the Chapter 16.

15.1 Possible components of the service offering

This section describes key service elements.

15.1.1 Plain old telephony (POTS)

POTS is simple voice service, which all WLL systems will provide as a minimum offering. The key concern here is voice quality. In all but the least developed markets, voice quality as good as conventional wireline will be essential.

15.1.2 ISDN

ISDN services were described in Section 3.1.2. Generally, ISDN is taken to mean data at rates of 144 Kbps or higher, providing at least two lines, either of which can be used for voice or data.

15.1.3 Fax

Fax service allows the transmission of documents using international fax standards, which are designed to operate over analog phone lines. If a digital phone line is provided, conversion will be required to ensure that the fax is transmitted correctly. Because most WLL vendors provide such capabilities within their system, both analog and digital WLL systems are likely to support fax.

15.1.4 Data

Data can be supported on analog lines with a modem and digital lines without a modem. The key issue is the data rate to be supported. Analog lines typically can support up to 33.6 Kbps. Digital lines can support up to whatever digital bandwidth is provided; for example, systems such as AirLoop can provide up to 384 Kbps and DECT up to 552 Kbps.

Most users will not see data as a service. They will have a particular application, such as the Internet, and will experience certain delays depending on the data capabilities of the system. In understanding the requirements for data, it is necessary to understand the use to which subscribers will put the system. That may entail understanding computer file transfer requirements and the data sent by some in-house packages.

15.1.5 Videophone

Videophone is a much talked-about service, which few providers, as yet, have adopted. It requires a data capability from the network of around 384 Kbps or higher, depending on the quality of the picture required; much lower bandwidths are possible but tend to result in pictures that move in a jerky fashion. If it is predicted that videophone will be widely used, relatively high bandwidth data links will need to be provided. When considering the requirement for videophones, it is important to remind users that the cost of a call will be substantially greater than the cost of a voice call.

15.1.6 Supplementary services

Supplementary services include a host of features, for example:

- Caller ID;
- Call-back-when-free;
- Call divert;
- Follow-me service;
- Advice of charge;
- Divert to voice mail;
- Call waiting;
- Barring of certain incoming and outgoing calls.

The list of possible supplementary services is almost endless. Supplementary services are supported in the switch, so if there is a requirement to upgrade supplementary services at a later date, the appropriate software can normally be installed on the switch. If the switch cannot handle the features, the cost per subscriber of replacing the switch is relatively low.

15.1.7 Centrex

Centrex is a service provided to businesses whereby instead of installing an office PABX, allowing, among other things, office workers to call each other using typically only three- or four-digit dialing codes, all calls are routed to the central switch, which acts like a PABX. The switch recognizes that the call has come from a particular office and decodes the short dialing number before routing the call to the appropriate person in the office. Centrex service provides savings for the company, which no longer needs to buy and maintain a PABX, and may offer better services. Although it provides increased revenue to the operator, it means that the operator needs to provide significant capacity on the air interface free of charge to route calls in a trombone from one user in the company back to another.

15.1.8 Operator services

Going one stage beyond than Centrex, some networks will provide an operator service, where callers phoning the switchboard number of a company are connected to an operator employed by the WLL operator and located central to a number of companies. The operator takes the call and forwards it as appropriate. That allows the company to reduce staff requirements, and because the central operator can act as a switchboard operator for a number of companies, staff efficiency can be improved.

15.1.9 Multiple lines

If a subscriber is provided with more than one line, then one line can be used for fax or data, without preventing the other line from being used for incoming or outgoing calls. Most residential customers do not now subscribe to multiple lines because the cost typically is twice that for a normal line. Offering multiple lines at a discount could be an appropriate marketing strategy. Most proprietary WLL technologies offer the facility of dual lines to subscribers.

Businesses, of course, will require numerous lines, typically 2 to 14, sometimes more. Systems capable of providing that capacity are available but typically are configured for the business market alone. High-capacity systems often are realized through the use of point-to-point microwave links in the 13- or 22-GHz band.

15.1.10 Leased lines

The concept of a leased line is a difficult one in WLL and worth exploring in more detail. The idea stems from the tariffing principles on existing systems. Generally speaking, telephony is tariffed on the basis of the amount of traffic generated. For users generating extremely high amounts of traffic, the PTO typically offers them a tariff where they can "lease" a dedicated line. They pay a monthly rental for the line (which is much greater than the monthly connection fee) but do not incur any call charges, regardless of how much information they send down the line. Users who lease a line are inclined to think they "own" the lines and that there should be no blocking on the lines. In practice, in a wired network, everyone has a dedicated line up to the switch, regardless of whether they decide to lease it or not; blocking then can occur in the switch on all the lines. Nevertheless, there is an expectation of a better service from a leased line than might be experienced from a nonleased line.

In a WLL network, the concept of a leased line is stretched even more thinly, because short of permanently reserving a slot on the air interface for a particular user, there is no concept of a dedicated resource that the user can "lease." Leasing simply becomes a different tariff package whereby the user elects to pay a high monthly fee and no call charges. Operators might like to offer such a package since it will accord with the manner in which large businesses currently pay for their communications. The difficulty arises if the user expects a better service as a result of leasing the line. In a typical WLL system, the same level of blocking is experienced by all users, so it will be difficult to differentiate between leased-line users and normal users. In such a system, if a better grade of service is required for the leased line users, they need to be on a different bearer.

A better solution is to make use of priority and preemption. The network would recognize that a leased-line user wished to access the network, would accord that user a higher priority, and would preempt the call of a lower-priority user to make the resources available. Operators will want to use such a tactic with care. Users become significantly more upset by a call that drops out part way through than by a failed attempt to start a call. If better service is offered to one user, it is possible that another user will become dissatisfied.

A compromise is to keep a number of channels free for higher priority users. When all the other channels are busy, a low-priority user will be blocked from the network when attempting to set up a call. High- priority users, however, will be able to access the spare channels. The number of channels reserved in that manner is a difficult choice: too high, and the air interface resources are badly used; too low, and high-priority users still will be blocked. Cellular operators have considerable expertise in setting that number because they tend to leave a pool of channels for subscribers handing over from nearby cells.

In practice, few of the existing WLL technologies offer priority and preemption, so hence this option rarely will be available.

15.1.11 Internet service provision

As well as having a line offering data capabilities, Internet users will need to have a subscription with an Internet service provider that provides their gateway into the Internet. A WLL operator could provide that value-added service to the customer.

15.1.12 Long-distance and international services

A WLL operator, in the strict sense of the word, owns the infrastructure connecting the subscribers' homes to the nearest switch. The revenue is associated with carrying telephony traffic across that segment. Whether the traffic is local or international is irrelevant, the operator receives the same fee for carrying the traffic. In practice, most WLL operators also own a switch. They interconnect the switch to long-distance operators and to international gateways. When a subscriber makes a long-distance call, the WLL operator routes the call from its own switch into the long-distance network. The operator charges the user for a long-distance call but has to pay interconnect fees to the long-distance operator, with the result that the WLL operator typically gains little more revenue for a long-distance call than from a local call.

There are a number of possible ways to increase revenue. One is for the WLL operator to deploy a long-distance network (i.e., connections between main cities). Such a deployment would be a trunked backbone network and is beyond the scope of this book. The simpler alternative is to negotiate an attractive interconnect agreement with the long-distance operator, allowing the WLL operator to keep a larger share of the revenue from the long-distance traffic.

Interconnect negotiations are beyond the scope of this book. There is little science to an interconnect negotiation but considerable negotiating skill and the need to understand the cost structure of the long-distance carrier. Such negotiations are best left to experts who have the detailed experience of having conducted them. The importance of such negotiations should not be underestimated—a favorable interconnect agreement can make a significant difference to the network profitability.

15.2 Mobility in the local loop

Most WLL networks have envisaged using fixed subscriber units mounted on the outside wall of subscribers' premises with a lead connecting to a phone socket inside the house. Radio, however, can provide a capability that copper cannot: mobility. Recently, a number of manufacturers and operators started to examine whether mobility should be provided in a WLL system. The reasons for considering mobility are varied. The fact that those involved in WLL typically have been involved in cellular radio certainly has had an impact. Manufacturers, particularly those using cordless technology, have stressed the added bonus of mobility in their WLL product literature. Literature about early WLL systems recounts stories about how the system performed well as a mobile system during some disaster. This section sets out the issues.

First, it is important to define mobility:

- Full mobility is the mobility offered by cellular radio, in that the user can expect to roam to most parts of the country and make and receive calls from a handset. Implicit in this definition is the handover of calls.

- Limited mobility is the user being able to move within a small area around the home, perhaps a radius of 1 to 2 km, and still make and receive calls. Call handover is not expected.

- No mobility is defined as a phone fixed by a wire into a base unit.

The provision of mobility has the potential to add significantly to the cost of a WLL system, due to the cost of enhanced coverage and the need for additional elements in the network.

A significant enhancement in coverage is required above that for a standard WLL deployment if mobility is to be offered. Compared to mobile systems, WLL deployments achieve a gain in link budget of up to 20 dB through the use of directional antennas mounted at roof-top level. To provide mobility in the home, a building penetration loss must also be built into the calculations; measurements have shown that for a reasonably reliable house penetration a further 20-dB loss must be allowed. The requirement for up to an additional 40 dB on the link budget has significant implications for the size of the cells. Use of propagation models applicable to WLL suggests that a 40-dB difference would reduce a cell size of 5 km without mobility to 400m with mobility. Experience with those deploying DECT systems has shown an even more severe coverage restriction, from 5 km to 200m when mobility is required. In the case of DECT, 600 times more cells are required if a mobility service is to be offered over the same coverage area.

For full mobility, a number of additional network components are required, including home- and visitor-location registers and gateway switches. Functionality, such as idle mode cell selection and handover, is required in the mobile, and the network and the cell planning must provide sufficient overlap to allow handover to occur. For some of the WLL technologies that have been derived from cellular, that functionality will be mostly in place. For others, it will not be possible to achieve the functionality.

For limited mobility, very little in the way of additional components is required, but the network planning may need to take care of propagation phenomena such as multipath, which would not be so severe in a deployment where there is no mobility.

The major issue, though, is the coverage. For in-building coverage, between 150 and 600 times as many cells will be required. Even if mobility is restricted to outside the house, allowing the range to increase to perhaps 1.4 km, 12 to 15 times as many cells will be required. They will add significantly to the network cost, which, as a crude approximation, varies linearly with the number of cell sites.

The communications marketplace is a competitive one, where the PTO, cellular operators, and, in a few cities, cordless operators increasingly compete for traffic. The WLL operators will be entering into that competitive environment and need to ensure that their offerings are targeted appropriately against the different forms of service that already exist. Table 15.1 summarizes the competitive environment.

WLL, as a replacement for fixed-loop telephony, needs to offer equally good services, including excellent voice quality (requiring a low BER channel), potentially ISDN, and services such as call-back, call forwarding, and follow-me. To what extent it needs to provide mobility and what price it can charge is the subject of this section.

It is clear from Table 15.1 that there is a strong relationship between price and mobility. Users are prepared to pay more and tolerate a lower grade of service to obtain mobility. The WLL operator potentially can

Table 15.1
The Current Competitive Environment

Operator	Mobility	Quality	Price	Services	Market
PTO	None	Excellent	Low	Wide ranging	All homes
Cellular	Full	Variable	High	Limited	All individuals
Cordless	Limited	Good to excellent	Medium	Limited	City workers and dwellers
WLL	?	Needs to be excellent	?	Needs to be wide ranging	All homes in coverage area

charge more for a service with mobility but cannot reduce the quality of the service offered below that of wireline. If prices increase while quality decreases, users will retain their wireline phones and the WLL operator will become a mobile operator using inappropriate equipment in an inappropriate frequency band.

This analysis allows the case for mobility to be stated quite simply. An operator needs to judge whether the additional cost it will incur in providing a service able to offer mobility can be offset by the greater call charges it can achieve, while still providing a service of equivalent quality to wireline.

A first pass at some of these figures would be as follows. The increase in network cost was calculated to be at least a factor-of-12 increase in the number of base stations, probably requiring a tenfold increase in the network cost. A comparison of cellular tariffs to PTO tariffs reveals that monthly rental is about 50% higher and call charges perhaps 100% higher (depending on a complex mix of time of day and bundling package). It is unlikely that the WLL operator will be able to achieve cellular revenue. Thus, the case for full mobility in the local loop simply can be discounted.

Indeed, a moment's lateral thinking helps validate those figures. Were it not expensive to provide indoor coverage, the cellular operators would be offering a WLL service on the back of their existing infrastructure. However, as One-2-One discovered in the United Kingdom, such a service is too expensive to provide and the revenues too low to justify the expense, even when the premium for full mobility was included.

There are, however, a number of options for offering different types of mobility that are considerably less expensive and likely to be of interest to customers.

- Provide mobility only in areas close to the base station. Even if a base station covers a radius of 5 km, those users within, say, 300m of the base station could have a mobile service around their houses and gardens. For no additional cost, limited mobility could be offered to a subset of subscribers, maximizing revenue. The difficulty in such an approach will be in determining which subscribers can receive the service and then marketing a range of different packages to different subscribers, depending on location.

■ Let users install cordless phones. An alternative would be to install a WLL line without mobility (which could be any technology) and then sell users DECT cordless home base stations. Such a phone and base station costs around $200 and will provide the user with mobility within a range of 200m from the house. The WLL operator could then choose to deploy DECT base stations in a few key locations, such as shopping malls, and then market a sort of cordless service.

■ Use dual-mode phones. Dual-mode DECT/GSM phones are being used on a trial basis and may become available during 1998. Some predict that by 1999, the price of dual-mode phones is expected to be only slightly greater than the price of a GSM phone, although others feel that the additional costs over a GSM phone will be much greater and that GSM itself will be adapted to provide cordless capabilities. Whatever, by selling the customer a WLL line to the house, a DECT or GSM home base station, and either a dual-mode DECT/GSM handset or an enhanced GSM handset in a bundled package, full mobility could be provided at no cost to the WLL operator. A user with such a phone could roam throughout the world. On returning home, the phone would automatically switch to the home base station when within range, routing calls across the WLL system. (Users could, of course, put together such a package for themselves, but with appropriate marketing the WLL operator could appear to be providing something different.)

The disadvantage of the third approach is that the WLL operator will not carry the traffic when the subscriber is out of range of the home base station and so will lose revenue. But, as shown earlier, capturing that revenue requires a disproportionate expenditure.

Table 15.2 shows the positioning that a WLL operator must adopt to be able to operate profitably.

The WLL operator needs to take advantage of the fact that wireless connections to a building generally are less expensive than wired connections to be able to operate from a lower cost base than the PTO. That allows WLL operators to attract customers with a combination of lower prices and differentiated services. Providing mobility will cause prices to

Table 15.2
The Current Competitive Environment

Operator	Mobility	Quality	Price	Services	Market
PTO	None	Excellent	Low	Wide ranging	All homes
Cellular	Full	Variable	High	Limited	All individuals
Cordless	Limited	Good to excellent	Medium	Limited	City workers and dwellers
WLL	None (except via home base stations)	Needs to be excellent	Low	Needs to be wide ranging	All homes in coverage area

be raised, at which point WLL operators will be providing a more expensive and inferior service than the PTO and a generally inferior service than the cellular operator. However, that does not need to stop the WLL operator from marketing a mobile service by offering home base stations and dual-mode phones (although the PTO also could market a similar service).

The provision of WLL is a complex matter, with WLL networks being used for a range of tasks from provision of telecommunications to areas that never have had communications, through shortening waiting lists in Eastern European countries to providing competition to the PTO in developed countries. The different environments require a range of different approaches, different technologies, and different services that will be the subject of further papers. Despite the wide range of networks, none of them would benefit from the provision of mobility. A WLL operator that wants to offer mobility should seek a cellular license.

16

Developing the Business Case

THE BUSINESS CASE is probably the single most important document in the process of deploying the network. With decisions having been made as to the technology, the density of the cells, the interconnect method, and the market niche, the business case is the point where all those come together to determine whether the project will be successful.

As well as being vital internally, the business case will be shown to investors and, possibly in a shortened form, to the government as part of the license application.

16.1 The overall structure of the business case

In a general sense, the business case balances the total outgoings in the form of network capital and on-going cost with the total incomings in the form of subscriber revenue. Neither of these calculations is simple. This section examines some of the typical issues and problems that arise.

233

The key drivers underlying the spreadsheet come from estimates of subscriber numbers, traffic levels, and tariffs over the lifetime of the network. Such estimates are more of an art than a science. The starting point is the gathering of as much data as possible about the target country in terms of the following:

- The population and the area of the country;

- The GDP per head and the variation in GDP across the country;

- The current teledensity, number of telephone lines, and distribution of those lines;

- The number of businesses and the distribution of those businesses according to the number of people employed;

- The current costs and growth in takeup of services offered by the existing telecommunications operators;

- The levels of penetration achieved by other competing operators (e.g., cable operators) where such competition exists.

Study of those figures, particularly in comparison with other similar countries, can reveal potential openings in the market. For example, such a study of the United Kingdom would reveal that the takeup of ISDN services was much slower than in France and Germany, and that that probably was a result of the relatively high tariff charged by BT. Based on that, a potential market opening exists in the provision of ISDN services.

More detailed study would include interviews with a selection of potential users to try to gauge their willingness to try a new telecommunications provider and the key services that they would require to switch to a new provider. Continuing the United Kingdom example, potential ISDN users probably would require a price at least 15% below that of BT's coupled with the same reliability and quality as achieved from BT.

There then follows a somewhat iterative process. Based on a service and price requirement and expectations of traffic figures, a preliminary network costing and financial modeling process can be performed. That will show whether the required services can be offered profitably. Depending on the result, it then may be necessary to increase the price, reduce the expected penetration, and rerun the model with the new variables until the most profitable operating position is reached. Assuming the model has been built on a flexible spreadsheet package, that need not be problematic.

16.2 The network build costs

The network build costs are divided into subscriber equipment costs and network equipment costs.

16.2.1 Subscriber equipment costs

The cost of a subscriber unit can be obtained from the manufacturer. There is likely to be a number of different subscriber units with different functionality and different costs, and each needs to be treated separately. The following information needs to be gathered:

- The projected number of new units required each year. That will be based on the predicted number of new subscribers expected each year. In the case where the subscriber units are owned by the operator, the operator can expect to reclaim most of the units from subscribers who leave the network; hence, the number of units required from the manufacturer will equal the net number of new subscribers per year. Otherwise, the number of units required will be approximately the gross number of new subscribers each year. (The gross subscriber number is the total number of new subscribers joining the network in the year. The net subscriber number is equal to the gross numbers minus the number of subscribers leaving the network.) In most cases, the projected number of new units per year over the network lifetime will follow the S-curve shown in Figure 16.1; hence, the spreadsheet will need a column for each year along with the actual number of new units expected during that year.

- The projected cost of the subscriber units over the investment period. In most cases, subscriber unit costs can be expected to fall as both manufacturing volumes and competition increase. The extent of such a fall is difficult to estimate because the market is embryonic; typical figures might be between 5% and 10% a year.

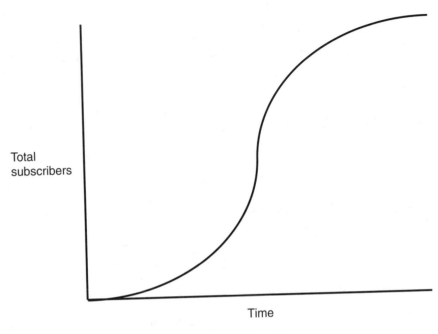

Figure 16.1 Idealized S-curve takeup of services.

Depending on the complexity of the spreadsheet required, either a maximum and minimum bounded approach can be followed, or an average of around 7.5% assumed. Discussions with manufacturers and potentially long-term supply contracts may set that annual cost reduction more precisely.

- Whether subscriber units are to be subsidized. If they are not, then it is not necessary to take their cost into account. However, most users expect to lease their facilities, so the network operator will need to purchase the units and recover their cost through rental over the investment period. In some cases, partial subsidy is used as an incentive for subscribers to remain on the network, which will have the effect of reducing prices. Typically, however, full subsidy will be used.

- The installation cost also will need to be included in the calculation. This cost is the average time taken to perform an installation multiplied by the labor cost for that time. Current estimates are that a two-person team typically can install four to five subscriber units a day. Thus, the installation cost for one unit is approximately half a worker-day in wages with associated administration overhead. That cost will vary from country to country. Other key variables will include the installation difficulty, which will be greater in less developed countries, and access to a trained workforce.

A typical spreadsheet for the calculation of the subscriber units' cost might then look like Figure 16.2.

Of course, projections are likely to continue for many more years than shown here. The spreadsheet has been restricted to five years so as not to overcomplicate the example. The key points for the analysis are that the cost of the subscriber units occurs gradually, building to a

Basic assumptions	
2-line unit	$500
8-line unit	$1,500
Installation	$150
Annual price reduction	7%
Subsidy	0%

Year	1998	1999	2000	2001	2002	2003
Units installed/ year (thousands)						
2-line units	10	30	60	100	100	60
8-line units	2	4	10	15	10	6
Cost						
2-line ($)	500	465	432	402	374	348
8-line ($)	1,500	1,395	1,297	1,207	1,122	1,044
Total cost ($millions)	9.8	24.6	49.4	75.6	65.1	37.0

Figure 16.2 Example spreadsheet for subscriber unit calculations.

maximum some time during the middle of the investment period in the network. That gradual spend is helpful in reducing overall investment requirements.

16.2.2 Network costs

The key components of the network costs are the following:

- Base stations;

- Base station interconnection;

- Base station controllers (if required);

- Base station controller interconnection (if required);

- Switching costs;

- Operations, maintenance, and billing system costs.

16.2.2.1 Base station costs

Base stations account for the following costs:

- The hardware cost of a base station, as quoted by the manufacturer. That cost will need to be adjusted according to the number of "cards" required in the base station, which will depend on the predicted capacity. Allowance typically is not required for the reduction in base station cost over time, because most will be purchased at the start of network rollout.

- The installation cost for a base station, including the cost of erecting masts where required, connecting to power supplies, and other similar costs.

- The planning overhead associated with each base station, requiring perhaps one day of planning time per base station.

16.2.2.2 Base station interconnection costs

The base station interconnection (i.e., connecting the base station to the switch) cost depends on whether leased lines or microwave links are used (discussed in Chapter 17). In the case that leased lines are used, a one-time connection fee may be payable to the PTO, but all other costs will be on-going rather than capital. In the case that microwave links are required, there is the full capital cost of the link equipment as well as the installation cost, although the latter typically is minor when conducted as part of the base station installation.

16.2.2.3 Base station controller costs

Some WLL technologies require that the base stations be connected to base station controllers. In that case, the full hardware cost of the base station controller must be included as well as installation cost. Most networks co-site the base station controller with either one of the base stations or the switch; hence, there is typically little site rental cost associated with this element.

16.2.2.4 Base station controller interconnection costs

Broadly, the same comments apply here as for base station interconnection.

16.2.2.5 Switch costs

The switching cost elements can include the following:

- The hardware cost of the switch;
- The cost of a dedicated building in which to site the switch;
- The cost of redundant power supplies, including a generator;
- Redundant interconnection to prevent a single point of failure;
- Appropriate security measures.

16.2.2.6 Operations, maintenance, and billing system costs

These costs typically will be co-sited with the switch; thus, the only cost will be the purchasing costs of the system.

16.2.3 Combining the cost elements

The cost elements are combined with the number of cells required, as provided by the coverage plan and the network plan, as explained in Chapter 17, and the rollout planning showing which years the expenditure will be incurred. Figure 16.3 shows a typical spreadsheet for a simple system in which base station controllers are not required.

The spreadsheet in Figure 16.3 shows a typical deployment, in which most of the base stations are installed in the early years of the network so that coverage is gained as quickly as possible. A single switch is required in the early years, but as the traffic increases in later years an additional switch is required to provide sufficient capacity. With 300 base stations, most cities in a country the size of the United Kingdom might be expected to have some coverage, depending on the range of the technology selected.

Basic assumptions (prices in $thousands)

Base station cost	50
Installation	20
Planning	5

Base station interconnect via microwave links

Cost	70
Installation	10
Switch costs	1,000
Switch installation	1,000

Year	1998	1999	2000	2001	2002
Base stations installed	50	150	100	0	0
Switches installed	1	0	0	0	1
Base station total cost ($millions)	7.75	23.25	15.5	0	0
Switch total cost ($millions)	2	0	0	0	2
Total cost ($millions)	9.75	23.25	15.5	0	2

Figure 16.3 Example spreadsheet for network cost.

16.3 The on-going costs

Once the network is operational, a host of on-going costs arise, including the following:

- Rental. Site rental will vary dramatically from high sites in city centers to low sites in rural areas. It is also extremely country specific.

- Leased line costs. These costs (discussed in detail in Chapter 17) are dependent on the capacity, length, and the PTO.

- Maintenance costs. These costs vary depending on the technology selected. As an approximate first-pass estimate, the industry average for the mobile telecommunications industry is between 1% and 2.5% per year of the capital cost. That compares favorably with the 5% cost typical of a wired access network.

- Radio spectrum costs. These costs also vary dramatically. In some countries, there will be no cost payable to the regulator for the use of the radio spectrum. In other countries, an auction fee, payable upfront and so becoming part of the capital cost, may be levied. Finally, annual fees may be required.

- Subscriber management costs. These costs are associated with sending bills to subscribers and managing any problems they might have.

- General management costs. These costs include management salaries, the cost of headquarters buildings, power and water bills, fleet-vehicle costs, and auditors' and consultants' fees. As an approximation, such costs might amount to around 1% of total revenue.

- Marketing, sales, and customer retention costs. These costs include direct marketing and sales expenses, as well as the cost of incentives such as loyalty schemes. Depending on the competitive environment, such costs should be below 1% of total revenue.

Figure 16.4 is an example spreadsheet showing how some of these costs might be modeled.

Basic assumptions

Average site rental ($thousands/year)	5
Cost per leased lines ($thousands/year)	15
Maintenance (%)	2.5
Radio spectrum ($thousands/year)	100
General costs ($/subscriber/year)	10
Marketing costs ($/subscriber/year)	10

Year	1998	1999	2000	2001	2002
Background information					
Total subscribers (thousands)	12	46	116	231	341
Total sites	50	200	300	300	300
Total network cost ($millions)	9.75	33	48.5	48.5	50.5
Total subscriber unit cost ($millions)	9.8	34.4	83.9	159.4	224.5
On-going costs ($millions)					
Rental	0.25	1	1.5	1.5	1.5
Leased lines	0.75	3	4.5	4.5	4.5
Maintenance	0.5	1.7	3.3	5.2	6.9
Spectrum	0.1	0.1	0.1	0.1	0.1
General costs	0.12	0.46	1.16	2.31	3.41
Marketing costs	0.12	0.46	1.16	2.31	3.41
Total on-going costs ($millions)	1.8	6.7	11.7	15.9	19.8

Figure 16.4 Example spreadsheet showing on-going expenses.

In this highly approximate example, where the base stations have been interconnected by leased lines, the key cost, particularly in later years, is maintenance. It shows the importance of reducing the maintenance percentage from the perhaps somewhat high 2.5% assumed here to much lower levels.

16.4 Predicting revenue

The prediction of costs is simple compared to the difficulties and uncertainties associated with predicting revenue. At its simplest, the calculation of revenue is just total subscribers multiplied by average revenue per subscriber. Unfortunately, both figures are subject to a high degree of error.

Subscribers need to be divided into different categories, such as large businesses, medium businesses, small businesses, and residential, and the total numbers and revenue predicted for each category. Tariffing policies need to be consistent with competition and target subscribers. Subscriber numbers need to show the change in subscribers over time and the change in call usage over time as Internet and other similar uses increase. Revenue needs to take into account the fall in call charges over time in liberalized environments.

Other factors to include in the calculation are the interconnection costs, taxes, and bad debts. Interconnection costs relate to payments to the PTO when a call originates on the WLL network but terminates in the PTO's network. Those costs vary dramatically from country to country, typically depending on regulatory intervention, but might amount to as much as 50% of all call-related revenue.

Figure 16.5 is an extremely simplified example spreadsheet showing some appropriate calculations for residential customers. Similar spreadsheets would be required for small, medium, and large business customers and any other groupings thought to be relevant. Operators must remember that the difficulty here is obtaining trustworthy forecasts of subscriber numbers and call tariffs. Current figures, historical extrapolation, and the advice of expert observers all should be used in an attempt to make the figures as accurate as possible.

In reality, considerably more complex spreadsheets than that in Figure 16.5 would be required.

Figure 16.6 shows the division of costs in a real WLL network. As can be seen, the main cost elements are those for the subscriber units and the base stations, but in this case, the leased line costs also are significant. Administration, advertising, and maintenance form a smaller part of the budget.

Basic assumptions

Annual subscription/customer ($)	100
Fall in subscription/year	10%
Call minutes/subscriber/year	4,000
Cost/minute ($)	0.1
Rise in usage/year	10%
Fall in cost/minute/year	10%
Interconnect fees	20%
Bad debts	5%

Year	1998	1999	2000	2001	2002
Total subscribers (thousands)	10	40	100	200	300
Total subscriptions ($millions)	1	3.6	8	14	18
Call minutes (millions)	40	176	480	1,040	1,680
Cost/minute ($)	0.10	0.09	0.08	0.07	0.07
Less interconnect	20%	20%	20%	20%	20%
Less bad debts	5%	5%	5%	5%	5%
Total call revenue ($millions)	3.0	11.9	29.2	56.9	82.7
Total revenue ($millions)	4.0	15.5	37.2	70.9	100.7

Figure 16.5 Example spreadsheet for predicted revenue.

Figure 16.7 shows revenue for the same WLL network. The revenue has been split into connection fees, monthly subscription fees, and call charges. The call charges have been split further into the call charge element kept by the operator and the interconnection payments made to the long-distance carrier. It can be seen that 13% of all revenue is paid out to the long-distance operator. Of the revenue, the call component forms the major part. That will vary from one network to another, depending on the tariffs and the volume of calls made by the subscribers.

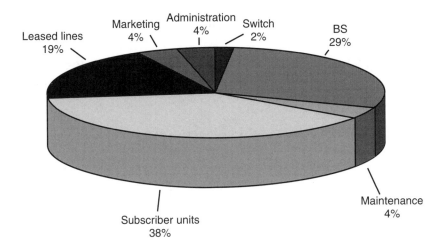

Figure 16.6 Division of costs in a real network.

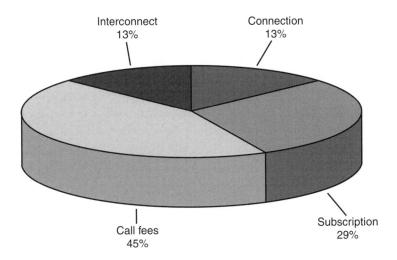

Figure 16.7 Division of revenue in a real network.

16.5 Financing arrangements

Even though WLL networks require relatively little upfront investment, compared to wired networks, financing is still required. This section shows how to calculate the total financing necessary. Two key issues arise in funding: how to get the capital and the interest rates that apply on the borrowing.

This section is not intended to be a detailed text on the ways of borrowing money; numerous financial texts are available on that topic. In outline, the following options are available:

- Self-funding from reserves within the group;

- Direct funding from banks or other financial institutions;

- Vendor financing from the equipment manufacturer;

- Shareholder funding.

The first funding option is the simplest because it does not require any direct action from the operator. However, the group will have an internal rate of return on capital that the group will want met or exceeded prior to the investment taking place.

Borrowing from banks has the advantage of simplicity. Typically there is no loss of control of the company, and repayment schedules can be set to meet the requirements of both parties. Banks, however, are wary of financing long-term and high-risk projects and typically only want to fund something like 25% of the capital requirements for a WLL network. If the total funding required is more than 25% of the total capital required, then banks may not present an appropriate route.

Vendor financing is much used at present. In return for buying the equipment from the manufacturer, the manufacturer is expected to provide funding by deferring the payments on the equipment. Manufacturers reluctantly agree to such projects if it is the only way in which they will sell equipment. However, vendors themselves can borrow only certain amounts of capital and cannot finance all the projects for which they are asked. Thus, an operator may have to select a vendor other than the one selling the most appropriate technology if vendor financing is required.

Shareholder funding requires the company to be floated on the stock market and that a number of investors be found. Shareholder funding has the advantage that there is little need for repayment schedules to be met but the disadvantage that some of the control of the company passes to the shareholders. The company now needs to meet targets and to operate successfully if it is not to risk being taken over.

For a more detailed introduction to financing methods, readers are referred to [1].

16.6 Summary financial statistics

With the total expected capital and on-going costs and the incoming revenue having been calculated, the next step is to compare the two to determine whether the selected balance of coverage, service, and tariffs meets the business requirements. There are many ways of examining the information; again readers who want a full introduction to project financial analysis should consult [1]. The first step is to compare the total expenditure with the total income, which is shown for the simple example in Figure 16.8.

Basic assumptions					
Interest payable on loans	10%				

Year	1998	1999	2000	2001	2002
Capital spend ($millions)	9.75	23.25	15.5	0	2
On-going spend ($millions)	1.8	6.8	11.9	16.2	20.2
Revenue ($millions)	4.0	15.5	37.2	70.9	100.7
Profit ($millions)	−7.6	−14.5	9.8	54.7	78.4
Borrowing ($millions)	−7.6	−22.9	−15.4	37.7	116.2
Interest payable ($millions)	−0.76	−2.29	−1.54	0.00	0.00
Bank balance ($millions)	−8.4	−25.2	−16.9	37.7	116.2

Figure 16.8 Example spreadsheet showing summary financial statistics.

The spreadsheet in Figure 16.8 shows an unusually profitable network (deliberately made so to keep the example simple); in practice, much longer break-even times are likely. Nevertheless, the spreadsheet adequately and simply demonstrates some of the key financial analysis that is required.

One key statistic immediately visible is the total funding requirement. The network will need to borrow over $25 million during 1999 before starting to repay the loans in following years. That is a key input into the funding analysis.

Another key figure is the time to break even. Figure 16.9 shows that the break-even point was achieved around three years after the network was constructed.

One of the most appropriate ways to evaluate a project is to consider the *net present value* (NPV). The NPV looks at the total outgoings and incomings over the length of the project, weighting money spent or received during future years by a factor accounting for the cost of capital over the period. Management texts agree that that is the most appropriate way of evaluating a project. To calculate the NPV, the table above needs

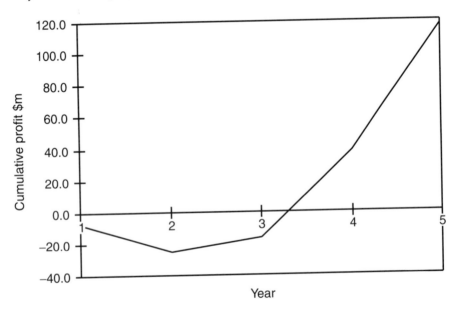

Figure 16.9 Cumulative profit by year.

to rearranged slightly to look at inflows and outflows and cost of capital, shown in Figure 16.10.

Figure 16.10 shows that when considered as a five-year project, the project has a NPV of $78.6 million. (In practice, most WLL networks will be considered over 7 to 10 years; 5 years was selected here for simplicity.) In principle, any project with a positive NPV should be undertaken.

Another widely used investment measure is the *internal rate of return* (IRR). The IRR effectively provides a measure of the interest that is gained on the money invested in the project. If the IRR is higher than the cost of borrowing the money or investing it elsewhere, the project should go ahead. IRR is widely disliked in management texts because of the following case. Say there are two mutually exclusive projects, one with an IRR of 20% and an investment of $100 million and the second one with an IRR of 30% and an investment of $20 million. There clearly is a much larger absolute amount of profit to be made on the first project, simply because of the larger sums involved. Because IRR does not consider the total sum invested, only the equivalent interest gained, then

Assumption	Discount factor	10%			
Year	1998	1999	2000	2001	2002
Discount	1.00	0.90	0.81	0.73	0.66
Total outgoings ($millions)	11.59	30.01	27.37	16.21	22.22
Discounted ($millions)	11.59	27.01	22.17	11.81	14.58
Total revenue ($millions)	4.00	15.48	37.16	70.86	100.67
Discounted ($millions)	4.00	13.93	30.10	51.66	66.05
Total ($millions)	−7.59	−13.08	7.93	39.84	51.47
Cumulative ($millions)	−7.59	−20.67	−12.75	27.10	78.57

Figure 16.10 Example spreadsheet showing NPV calculations.

a company would, incorrectly, select the second investment. For that reason, NPV is considered the more appropriate investment mechanism. However, because IRR is used so widely, it is included in the discussion here. IRR is, in fact, closely related to NPV analysis; the IRR is nothing other than the discount rate, which, if applied to an NPV analysis, would yield an NPV of exactly zero.

Calculation of IRR where payments and revenue take place over a number of years is, actually, extremely difficult to solve mathematically. The best approach is an iterative solution, in which different discount rates are entered into an NPV analysis until an NPV of zero is reached. Alternatively, spreadsheets are widely available with IRR macros that perform an iterative calculation automatically. Figure 16.11 shows an iterative approach to working out the IRR.

As can be seen, the IRR for this project is an extremely high 49.2%.

16.7 Tariffing policies

One key decision for the operator is the tariffing policy that will be used. Tariffing is a complex topic, one that is beyond the scope of this book. In one respect, the tariff must differ from that available on a wired system. Wired tariffs often include a call connection component, designed to make the subscriber use the channel for a minimum period, to maximize

Assumed Discount Factor (%)	Actual NPV ($millions)
10	78.6
40	9.2
60	−6.3
50	−0.68
48	0.92
49	0.1
49.2	0

Figure 16.11 Iterative approach to solving the IRR.

the revenue of the operator. That policy may be beneficial where there is a dedicated channel to the subscriber, but in the case of WLL systems, where blocking on the air interface is a key problem, it can have the effect of reducing revenue.

In the case of data transmission, it is possible that the data to be transmitted will arrive in bursts. If there is a connection charge or a minimum call charge, it may be less expensive for the user to hold the channel open for a period of time, in case there is any more data to send, rather than break the channel down and reestablish it when required. The effect of such a policy is to increase the cost to the user of sending data and results in wasteful use of air interface channels.

Reference

[1] McLaney, E. J., *Business Finance for Decision Makers*, Pitman, 1991.

17

Rolling Out the Network

THE OPERATOR has now reached the point of being ready to start deploying the network. The license has been granted, the frequencies are available, the technology has been selected, the finance has been arranged, and the business case describes the services and capacity to be offered. This chapter describes the process of actually getting a network up and running.

In many ways, this approach is similar to that followed by cellular operators, so those familiar with cellular deployment may want to skip this part.

An overview of the process of setting up the network is provided in Figure 17.1; a schematic of a WLL network is shown in Figure 17.2.

Each of the key steps in the block diagram in Figure 17.1 is described next, starting with selecting the number of cells, moving to choosing the cell sites, connecting the cells back into the network, and interfacing with switching and backbone systems.

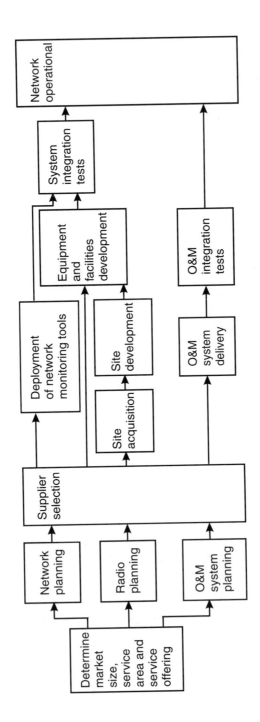

Figure 17.1 The network rollout process.

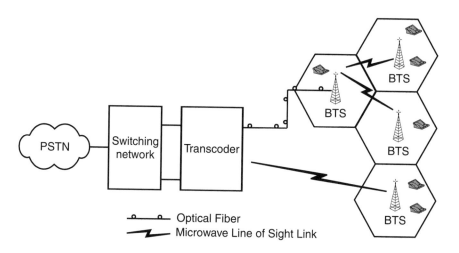

Figure 17.2 Schematic of a WLL network.

17.1 Selecting the number of cells

First, a few words on terminology. There often is confusion over the use of the terms *cell sites* and *base stations*. Most of the time, cells are the same as base stations. However, where cells are sectorized, some use the terminology that each sector is a base station (because it requires a separate transmitter). Such confusion is avoided here with this definition: a cell site is a single location with one or more sectors.

The total number of cell sites is a critical parameter for the network. It is one of the key cost drivers in setting the total network cost because the main element of the network cost is the number of cells. Too many cells and the network costs will be higher than necessary. Too few cells and there will not be adequate coverage or capacity in the system. Although it is possible to add additional cells later, optimal deployment might require resiting earlier cells, which will engender additional expense.

The number of cells is driven fundamentally by the business plan, as described in Chapter 16. The key inputs from the business plan are the following:

- The number of homes;
- The density of the homes;

- The expected penetration;
- The expected traffic per home.

Those inputs are used to determine how many cells are required to provide adequate capacity. The other important set of parameters is as follows:

- The area to be covered;
- The range of the system;
- The topography of the area.

That set of parameters is used to determine how many cells are required to provide adequate coverage. Once calculations have been made as to how many cells are required for capacity and how many for coverage, it is necessary to take the higher of the two numbers. Another way of looking at it is that it is necessary, at a minimum, to provide enough cells to cover the target area; if that does not provide sufficient capacity, more cells will be needed.

The calculation as to how many cells are required needs to be performed for each part of the country where conditions differ. That is, there is little point coming up with the required base station density for an entire state in one calculation if that state actually consists of one dense city and otherwise rural areas. The base station density would be too low to provide adequate capacity in the city and too high in the surrounding rural areas. The calculation should be performed for areas where any of the following factors differ significantly (say, by more than 20%) from another area:

- Density of homes;
- Expected penetration;
- Traffic per subscriber.

That might result, for example, in separate calculations being performed for the financial district of the city, which has high telecommunications demand, and a neighboring area of the city that might contain businesses with lower telecommunications demand. The need to perform the calculations numerous times is not problematic in itself; as will be seen, the calculation is relatively simple and could be computed quickly on a spreadsheet. The difficulty is in obtaining the input information. Here it is assumed that the information has been obtained as part of development of the business case.

The number of cells required for coverage in a particular area is given by

$$Number\ of\ cells = \frac{size\ of\ area(km^2)}{\pi r^2} \cdot I \qquad (17.1)$$

where r is the expected cell radius in kilometers, and I is a factor that represents the inefficiency of tessellating cells

The expected radius can be obtained from the manufacturer or from trial results. For those who want to understand this in more detail, a link budget can be constructed and the required path loss linked to the cell radius using propagation modeling.

The inefficiency factor derives from the fact that cells are not perfect hexagons, and it is not always possible to select cell sites exactly where a plan would require them to be. As a result, coverage areas from neighboring cells often overlap (if they do not, there is a gap in the coverage, which may need to be filled with an additional cell). The size of the inefficiency factor varies, depending on the topography, the skill of the network planner, and the availability of plentiful cell sites. As a guideline, a factor of 1.5 (i.e., a 50% increase in the number of cells) would not be uncommon, and a factor as high as 2 would be experienced in some situations.

The number of cells required for capacity is given by

$$Number\ of\ cells = \frac{traffic\ channels\ required}{traffic\ channels\ per\ cell} \qquad (17.2)$$

It is necessary to calculate both of those factors. The number of traffic channels required is given by

$$Number\ of\ channels = E[number\ of\ subscribers \cdot penetration\ (\%)$$

$$\cdot\ busy\ hour\ Erlangs\ per\ subscriber] \qquad (17.3)$$

where $E[x]$ represents the conversion from Erlangs to traffic channels using the Erlang formula. That is simply a formula that describes how many radio channels are required to ensure that blocking is no worse than required for a given amount of traffic. The Erlang B formula is given by

$$P_B = \frac{A^N/N!}{\sum\limits_{n=0}^{N} A^n/n!}$$

$$(17.4)$$

where P_B is the probability of blocking, A is the offered traffic in Erlangs, and N is the number of traffic channels available.

The number of traffic channels, N, required for a given A and P_B can only be solved iteratively, by substituting values of N in (17.4) until the desired P_B is reached. As a result, tables of Erlang formula are available. A graph showing the variation of radio channels required with Erlangs of traffic and different blocking probabilities is shown in Figure 17.3. Those who want to know more about the Erlang formula should consult [1].

Note that the slight steplike nature of the curves is caused by the fact that only an integer number of channels is possible, so the number of channels tends to step slightly. Note also that the number of channels tends toward the number of Erlangs for high offered traffic (e.g., for 1 Erlang, 5 channels are required, an efficiency of 20%, whereas for 19 Erlangs, 25 channels are required, an efficiency of 76%). That phenomenon is known as trunking gain: the more traffic that can be trunked together, the more efficient the system.

17.2 Selecting the cell sites

Having decided on the density of cells, it is then important to select the best cell sites. Broadly speaking, if the system is coverage constrained,

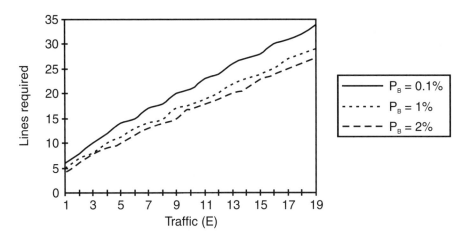

Figure 17.3 Channels versus Erlangs for a range of blocking probabilities.

then cell sites on high buildings or high areas will be important. If the system is capacity constrained, then lower buildings may be more appropriate because they will allow other buildings to act as shields, dividing cells and preventing the signal from spreading too far.

As a first pass, the coverage area should be divided into a regular grid (normally a hexagonal grid), with the radius of the hexagons being the calculated cell radius as required for coverage or capacity. Base station sites then should be sought as near as possible to the centers of the nominal hexagons.

The next stage is to make use of an appropriate planning tool, as described in Section 6.6. With the use of a planning tool, base station sites can be tried near the center of the nominal hexagon until one is found that approximates as closely as possible to the desired coverage. The process is repeated, building up the entire grid. As one base station is selected, neighboring cells may need to move increasingly from the nominal grid, such that they provide a minimal overlap with the first cells.

Once the sites have been selected on the computer, they should be checked visually to make sure that the LOS paths are as predicted. Suitable mounting points for antennas should be investigated, as should equipment housing and power supplies. If all these look promising, it is time to enter into negotiation with the owner of the building for the rental of space to site the equipment. There is little that can be said about this, other than a good negotiator may be required and it might be helpful to find out the going rate for rental from other operators.

One approach that can avoid the need for site negotiation for every cell is to make a deal with an organization that owns numerous buildings, such as a bank. If that is done, the radio planning becomes quite different. Instead of working from a grid, base stations are now simulated on all the banks (or whatever), and sites then are sought to fill in the resulting gaps. The number of cells required by this approach should be compared with the number required if the buildings to be used are unconstrained. If additional cells are required when using the banks, the extra cost associated with the additional cells can be compared with the savings made for site rental and time spent negotiating to determine whether it is cost effective to use sites on the banks.

Radio planning is a resource-intensive procedure. Trained operators will need to spend some time using planning tools, trying different configurations and cell sites to achieve the best layout. Depending on the complexity of the environment, the number of cells planned per person per day can vary dramatically. Operators may consider using contract staff to perform the planning; as once the network has been rolled out, there will be little need for such staff.

There are some subtleties to the cell site planning. The first is the use of sectorization. Instead of using an omnidirectional antenna providing coverage over a circle, directional antennas are deployed, splitting the coverage into a number of wedge-shaped slices. Nonsectorized and sectorized cells are shown diagrammatically in Figures 17.4 and 17.5, respectively.

The benefits of sectorization differ dramatically for CDMA and TDMA systems. For CDMA, sectorizing a cell results in an increase in capacity, broadly in line with the number of sectors deployed. That is because adjacent cells already use the same frequency. By splitting a cell into a number of sectors, each sector can use the same frequency. The result is that the interference to adjacent cells is increased, because there are now more users in the cell that has been sectorized. That reduces the capacity in the adjacent cell, but the reduction in capacity is much less than the gain in capacity in the cell that has been sectorized. To a first approximation, the capacity is increased in line with the increase in the

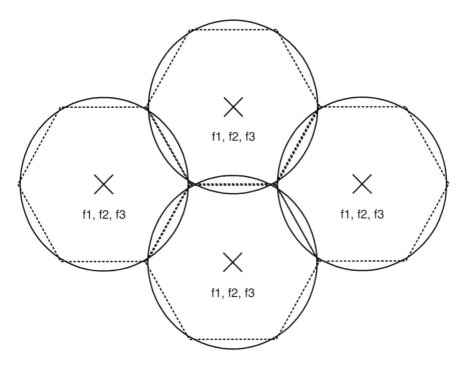

Figure 17.4 Omnidirectional cell arrangement.

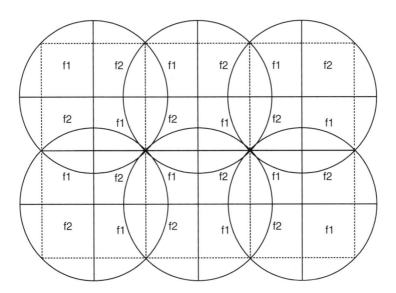

Figure 17.5 4-sector cell arrangement.

number of sectors (e.g., when moving from a nonsectored to a three-sectored cell, the capacity increases by a factor of 3).

For TDMA, the gains are much more limited. Because the base station still transmits with the same power after sectorization, neighboring cells are unable to use the same frequencies. And because additional frequencies are required in the sectorized cell (since each sector must use a different frequency), the overall number of frequencies required and, hence, the cluster size, rise. That has the effect of offsetting the gains in capacity associated with making the sectors smaller than the original cell. Careful planning using the directionality of the antennas in the sectors can result in some gains. For example, moving from a nonsectored system with a cluster size of 3, then a three-sectored system would require a cluster size of around 7 or 8. The overall gain from sectorization is then around between $(3 \times 3)/8 = 1.1$ to $(3 \times 3)/7 = 1.3$. With dynamic channel allocation, it may be possible to increase those gains slightly more.

Each sector requires a separate base station unit, resulting in additional cost, but the site rental typically is unchanged, making the use of sectors on a single cell less costly than installing multiple small cells.

Given the limited gains of sectorization for TDMA systems, it is reasonable to ask why most GSM operators use a sectorized cell system in city areas. The reason is that sectorization allows for a greater path loss than a nonsectored approach because the gain of the sectored antennas can be included in the link budget. That has the effect of increasing the range. Given that a number of base station transceivers would be required to handle the traffic generated in city areas anyway, connecting those transceivers to different antennas results in little increase in cost. Further, in city areas, sectorization prevents some multipath reflections that might occur if an omnidirectional antenna was used, increasing signal strength and finally a small capacity gain is achieved. Overall, for a small cost increase, improvements in capacity, range, and signal quality can be achieved by sectorization in this area.

In suburban and rural areas, however, sectorization is used less. That is because in those areas it may be that only one carrier is required to provide sufficient capacity. To use additional carriers to feed each sector would engender a significant cost increase.

Another approach is the use of hierarchical cell structures. In a hierarchical structure, an oversailing macrocell provides coverage to a large area, which typically generates more traffic than the capacity provided by the cell. Smaller microcells or minicells then are inserted in areas of particularly high traffic density, taking some load off the macrocell. Such minicells need to use a different radio frequency to avoid interference. That can be a sensible approach in a large city where the average traffic levels are low, but there is a small area in the center with high traffic densities. If the nonhierarchical approach is adopted, then one cell would be required in the center and perhaps six cells around it to cover the entire city. Each of the six cells would be operating at well below capacity. With a macrocell-minicell approach, only two cells provide coverage in the same area.

17.3 Connecting the cells to the switch

Once the cells have been selected and the cell sites commissioned, it is necessary to connect the cells to the network, typically to the switch (possibly through a base station controller unit, depending on the architecture of the technology selected). There are three general ways to do that:

- Using a link leased from the PTO;
- Using a microwave point-to-point link;
- Using a satellite link.

Important factors in making the decision include the following:

- The relative cost, which is affected by the distance of the link, the capacity required from the link, the presence of available infrastructure, and the availability of radio spectrum;
- The desire to avoid using a link provided by the PTO, which probably is a competitor;
- The desire to standardize on the interconnection method throughout the network.

Each of the interconnection systems is explained in more detail in the following sections.

17.3.1 Leased link

Leased link is the leasing of a line from the PTO. The line may not actually be cable—the PTO itself may be using satellite or microwave links—but that will be transparent to the operator. The operator pays an annual fee for use of the link, which typically is related to the distance over which the link runs and the capacity required from the link. There also may be an initial connection fee to link the base station to the nearest access point in the leased link network. PTOs typically publish books of prices for such services.

Even when the leased link is the cheapest option, operators may want to avoid it for a number of reasons. Links may not be sufficiently reliable or have a sufficiently low error rate. The operator, for competitive or whatever reasons, may want to avoid leasing key resources from a competitor they might suspect of deliberately providing an inferior service to make the competition they suffer from the WLL operator less severe.

Table 17.1 shows some leased costs charged by the PTO in the United Kingdom at the beginning of 1997.

Table 17.1
Example Fixed-link Costs

Link Capacity		2 Mbps	34 Mbps
Connection costs	First link	$18,800	$94,000
	Additional links	$5,900	$47,000
Annual Rental	Both ends in London	$7,700	$77,000
	Else		
	If link <15 km	$11,000+	$110,600+
		$600/km	$6,100/km
	If link >15 km	$16,000+	$202,000+
		$300/additional km	$3,000/additional km

17.3.2 Microwave links

Microwave links are point-to-point radio devices. A radio transmitter and a directional antenna are placed at the cell site, pointing to another directional antenna and a receiver at the switch site. Microwave links operate in a number of frequency bands, with differing ranges. Figure 17.6 shows how the expected range of the microwave link varies with distance.

For distances greater than can be achieved with the available frequencies, additional hops may be required, adding to the expense. Further, spectrum assignments, particularly in the lower frequency bands, may not be available. As an example of that, Ionica, a WLL operator in the United Kingdom, has stated that it would have liked to interconnect all its base stations to its switches using microwave links at 7.5 GHz. However, due to spectrum congestion in the United Kingdom, spectrum was available only to satisfy 10% of the requested links. Other links had to be at higher frequencies or using leased cable.

Some excellent texts are available on the difficulties associated with deploying fixed links, for example, [2]. Broadly, the same LOS radio propagation problems as described for WLL in Chapter 6 occur (the radio path from a WLL base station to a single subscriber looks similar to a microwave fixed-link radio path).

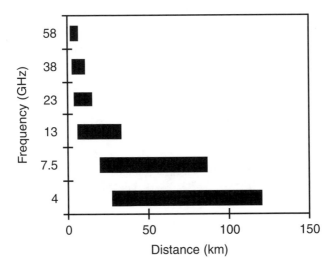

Figure 17.6 Microwave link range variation with frequency.

The cost of microwave links lies in the cost of the link equipment, site rental (although that may not be required if the cell sites can be used), and ongoing maintenance. Table 17.2 shows typical values for some of those costs.

In the reading of Table 17.2, it is important to remember that annual site rentals and license fees are country specific; the examples here are for the United Kingdom. The maintenance level will depend on the local environment; here 2.5% per annum of the equipment cost has been assumed. The total cost is for a 10-year deployment, and the discounted cost takes into account a cost of capital at 10%. Equipment in the 38-GHz band is particularly inexpensive because of the high economies of scale generated in that band by the use of those frequencies by cellular operators for interconnecting base station sites to switches. The site rental cost falls with higher frequencies because smaller antennas are used that place a lower wind loading on the mast. License fees also fall with higher frequencies to reflect the lower congestion in those frequencies.

The cost of a fixed link generally is not related to either the distance of the link or the capacity required. Hence, when fixed-link costs are compared with leased-line costs, the result typically is that shown in Figures 17.7 and 17.8, for 2 Mbps and 34 Mbps links, respectively.

The cost of leased lines in each country varies, so it is not possible to determine where the crossover occurs. As a general guide, fixed links are not economic for bandwidths below 2 Mbps or distances below 15 km. For most cell site–to–switch interconnects, bandwidths of around

Table 17.2

Typical Costs ($) in 1997 for Fixed Links in Various Frequency Bands

Frequency	Equipment Cost	Annual Site Rental	License Fee	Annual Maintenance	Total Cost	Discounted Cost
4 GHz	80,000	4,800	1,600	2,000	164,000	136,280
13 GHz	80,000	4,800	1,600	2,000	164,000	136,280
22 GHz	60,000	3,000	1,000	1,500	115,000	96,850
38 GHz	48,000	2,400	800	1,200	92,000	77,480
55 GHz	60,000	2,400	400	1,500	103,000	88,810

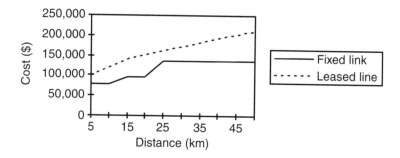

Figure 17.7 Fixed-link versus leased-line costs for a 2-Mbps link.

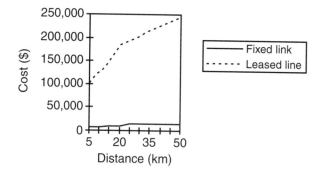

Figure 17.8 Fixed-link versus leased-line costs for a 34-Mbps link.

2 Mbps or higher will be required, so microwave links may be an economic alternative.

Microwave link cost is increased if high reliability is required. To achieve higher reliability, additional radio units are placed on the same mast in a "hot-standby" mode, so they can be switched on immediately if the original unit fails. That doubles the equipment cost.

In considering the reliability required, it is important to understand whether links have been daisy-chained. Examples of a daisy-chained and a non-daisy-chained arrangement are shown in Figure 17.9: signals from some cells are carried over a number of fixed links, increasing the need for reliability.

Simple probability modeling can be used to determine whether redundancy is required based on reliability figures quoted by the manufacturer, the reliability required by the business case, and the network

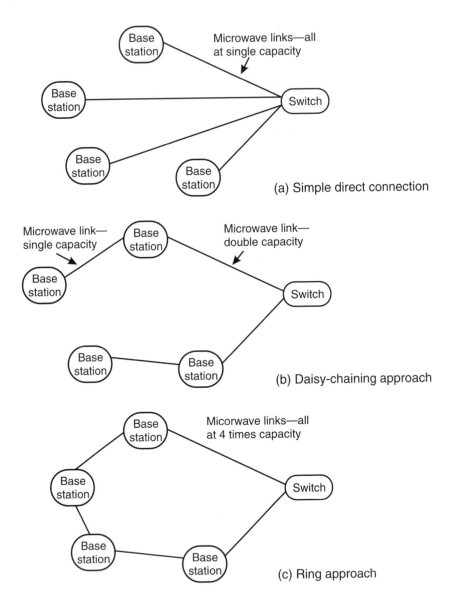

Figure 17.9 Various interconnection arrangements.

model. For example, if each link has a reliability of 99.9%, then for any of the base stations in arrangement (a) shown in Figure 17.9, that is the reliability experienced. In arrangement (b), for the two base stations linked to other base stations, the reliability is $0.999^2 = 99.8\%$. For arrangement (c), for a base station farthest from the switch, one of the links on the clockwise and anticlockwise route back to the switch needs to fail simultaneously. The probability of the clockwise branch not failing is 99.8%; when it has failed, there is a 99.8% probability that the other link has not failed. Hence, the overall reliability = 0.998 + 0.002 × 0.998 = 0.999996, or 99.9996%. Clearly, the ring approach provides significant reliability gains.

Figure 17.9 also shows that those gains come at the cost of an increase in the link capacity. In the case of a ring, each link on the ring must be capable of carrying something close to the total capacity required by all the base stations on the ring so that if one branch fails, all the traffic can be passed through the remaining branches.[1]

17.3.3 Satellite links

With the use of *very small aperture terminals* (VSATs), base stations can be linked to the switch via a satellite link. Typically, satellite links are more expensive than fixed links and cause a significant delay to the signal, which is perceived as delay or echo by the user and which is highly undesirable. However, in situations where the cell site is so remote from the switch that a number of hops on a fixed link would be required and there are no available leased lines, satellite links may form the only viable alternative.

To establish a satellite link, capacity must be leased from a satellite operator, such as Inmarsat, which will provide assistance in installing the dishes and appropriate transmitter and receiver equipment. The cost of a 2-Mbps leased line over satellite currently is around $480,000 per year, depending on location, so a 10-year, through-life cost would be around $3.2 million, substantially more expensive than the other options.

17.3.4 Protocols used for the interconnection

Clearly the base station and the switch need to use the same protocol along whatever medium is used to interconnect them. The protocol

---

1. To be more precise, those links at either end of the ring must be capable of carrying the total ring capacity; the one in the middle, half the link capacity; and those in between need a capacity increasing linearly between the center and the ends. That can be easily seen if it is imagined that either one of the links nearest the switch fails.

describes how speech and data calls are carried on the line and sends signaling such as establish call and disconnect call. If the switch and the base stations are purchased from the same manufacturer, the protocol may be irrelevant—it can be assumed that the manufacturer will provide equipment that will interconnect. The key issue is that some manufacturers have protocols that are efficient at compressing, or concentrating, the signals to be transmitted, reducing the bandwidth requirements on the link to the switch. Clearly, the more the signal can be concentrated, the more the cost can be reduced, particularly if leased lines are being used.

All those protocols run over E1 data links which are 2-Mbps (2.048 Mbps to be precise) links. The link is segmented into thirty-two 64-Kbps slots. Slots 0 and 16 are reserved for control data and synchronization information; hence, each E1 bearer can provide up to thirty 64-Kbps channels.

The following main protocols are available:

- *Channel-associated signaling* (CAS);

- *Signaling system 7* (SS7);

- Q.931;

- V5.1;

- V5.2.

CAS can carry only POTS (simple voice) traffic, while Q.931 can carry only ISDN (data) traffic. CAS suffers from one key disadvantage: it is what is termed nonconcentrating. That means that for every subscriber, a time slot on the E1 bearer must be permanently reserved, regardless of whether a call for that subscriber currently is in progress. Because most subscribers generate traffic only around 5% of the time, that is extremely inefficient. Its only benefit is that blocking is not possible on the base station-to-switch interface.

The V5.1 protocol also is nonconcentrating, but it has the advantage of being able to carry both POTS and ISDN traffic. That is a significant advantage, as the following example demonstrates. In a network where both POTS and ISDN are provided (as will be the case for many WLL networks), if CAS and Q.931 are used, then separate E1 lines would be required for CAS signaling and Q.931 signaling. If each required only half the capacity of an E1 line, then without V5.1, two E1 lines would be required, whereas with V5.1, only a single line would be required.

The V5.2 protocol is concentrating, so it is not necessary to dedicate resources on the E1 link to any particular subscriber—they can be assigned dynamically (in the same fashion as is performed on the air interface). This allows far fewer E1 links to be used, reducing the overall network cost.

At the beginning of 1997, the V5.2 standard appeared to be becoming widely accepted as the international means of connecting switches. Most manufacturers of WLL networks were in the process of developing a V5.2 interface for their products, due for release around the beginning of 1998. However, there was also significant concern that there were many options in the way that a V5.2 interface could be configured and that V5.2 interfaces from different manufacturers would not necessarily interwork. That matter needs to be monitored closely over the coming year, but at this point, V5.2 would appear to represent the best solution to interconnection.

If V5.2 is not used, it is possible that the system capacity is limited not by the air-interface capacity, but by the number of subscribers that can be accommodated on the E1 links to the switch. For example, in one system, the base station allowed a maximum of 16 E1 connections back to the switch using a V5.1 protocol. Given that each connection could support 30 subscribers, the total cell capacity was limited to 480 subscribers. However, in a fully loaded system where the subscriber traffic is assumed to be low (0.1E) and the blocking tolerated high (1%), the air interface capacity could be as high as 850 subscribers. Therefore, the main point of blocking will be the switch connections, not the air interface.

In such a situation, some strange decisions are required. For example, rather than users being given a 2B + D ISDN connection, requiring two (or more) slots on the E1 interface, they would be given a B + D connection requiring only a single slot. They could transmit the same total amount of data, but it would take twice as long. The air interface traffic would be unchanged (just less bursty), while the E1 resourcing would be eased. Clearly, the designer of a network does not want to be placed in such a position.

17.4 Installing the subscriber units

Once the network is operational, it is necessary to install subscriber terminals. The difficulty associated with that depends on the network design and the subscribers' residences. In the simplest case, the signal is sufficiently strong inside the house that no external antenna is required. The phone, with an antenna attached, is screwed to the room in the house where it will be used. Apart from some simple tests to check that there is adequate signal strength, such installation is straightforward.

Most WLL systems do not operate in that fashion. They require an external antenna mounted at rooftop level or on the side of the building to ensure a high-quality link of sufficient range. Installation is a case of bolting the antenna to the side of the house and then running a cable, normally down the outside of the house and then into the room where the phone socket will be installed. The socket then is installed in the room and the phone connected.

The skill is in mounting the antenna in a location where there is a good signal and orienting the antenna in the correct direction. To do that, the installer requires some training. On arriving at the house, the installers need to know the direction of the base station, information that can be provided on the job card by the network planning tool. They need to look in that direction from the house and take note of possible obstructions, such as trees (remembering that trees in full leaf have a greater attenuation than bare trees), and looking for points on the house where such obstructions can be avoided or minimized. Having decided on the approximate location on the house, the installers need to make signal strength measurements, slowly moving the measurement device to ensure that it is not situated in a fade. If the signal strength is satisfactory, the antenna unit can be mounted, the antenna aligned to maximize signal strength, and the remainder of the installation completed. If there is inadequate signal strength, that should be reported to the network planner so the reason can be ascertained.

In some parts of the world where WLL is deployed, the houses are too low or too unstable for subscriber units to be mounted. Shantytowns are a good example of such locations. The normal solution is to mount the subscriber unit on a pole erected specially for the purpose and situated alongside the building in which coverage is required. That approach clearly is more costly than mounting directly on the house but often it cannot be avoided. The poles need to be sturdy; they then can be used by the shantytown dwellers as something they can bolt their houses to increase stability.

17.5 Billing, customer care, and related issues

Generally, there is little that is new here. Billing systems can be taken from either fixed or mobile networks and used directly in a WLL environment. Customer care help lines will be very similar to those implemented by the fixed operator. Perhaps the key issue for WLL is the requirement to enter the subscriber's premises and bolt equipment onto the house. That can be problematic for a number of reasons:

- Subscribers may not be home at the time of the appointment, resulting in a wasted visit and the additional cost of a later visit. That situation generally can be avoided if the subscriber can be given an exact time that the installer will arrive, if visits can be scheduled for evenings and weekends, and if the subscriber is called a few hours before the visit to confirm the installation.

- The installers may be the only persons employed by the operator whom the subscribers meet. It is important that the installers be well presented and generally give a good impression of the operator.

- Subscribers may not be prepared to let the installers drill the necessary holes in their house; in particular, running the cable into the living room often is problematic because it may affect interior decorations. Installers need to be well trained at making a minimum of disruption, making good afterward, and being inventive in looking for alternatives that might be more acceptable to the subscriber.

17.6 Summary

By now, you should understand a good deal about the concepts and ideas that make up WLL. Going from an understanding of concepts to putting the ideas into practice is, of course, a large step. Remember to follow these steps:

1. Identify the target market.
2. Determine the service offering required.
3. Build a detailed business case.
4. Select the technology.
5. Apply for a license.
6. Gain appropriate financing.
7. Design and build the network.
8. Market the service to gain the required number of customers.

Also remember that expertise exists in the form of consultants and experienced individuals. Making the best use of those resources can reduce significantly the cost of becoming a WLL operator.

Chapter 18 illustrates some of these concepts through the use of a case study.

References

[1] Schwartz, M., *Telecommunication Networks—Protocols, Modelling and Analysis*, Reading, MA: Addison-Wesley, 1987.

[2] Freeman, R., *Radio System Design for Telecommunications (1–100 GHz)*, New York: Wiley & Sons, 1987.

18

Case Study

THIS CHAPTER PRESENTS a case study of the design of a WLL network to show how all the different concepts in the preceding chapters come together.

This case study is based on the design of a real WLL network for Colombia, concentrating mainly on the capital city, Bogota. To preserve commercially confidential information, some of the key market forecast figures and equipment parameters have been modified. Nevertheless, the principles illustrated by the case study are relevant regardless of the accuracy of the underlying information.

18.1 Market demand forecast

This section examines the predicted market demand. The market demand is critical in determining the spectrum and interconnect requirements for the system and in driving the business case. First, predictions as to the number of lines are presented, followed by predictions as to the traffic generated by the different subscribers and the likely destination for that traffic.

18.1.1 Provision of lines

The determination of the number of lines is based on two key inputs. The first is the total population of each of the key cities where a WLL network is to be deployed. The second is the expected penetration of each of those markets over time. Estimating the expected market penetration is difficult. In the case of Colombia, the following factors were taken into account:

- In most cities in the country, there currently are two providers, both using wired solutions. Only in Bogota are there any WLL operators. In Bogota, it is predicted that there will be two wired and three wireless operators over the network lifetime.

- Current penetration typically is 40% of homes in the key cities. The demographic information available shows that 40% of homes are too poor to be able to afford a telephone service over the network lifetime. Overall penetration might reach 60% of homes (but a much lower percentage of the population) over the network lifetime.

- In most cities, the existing operators provide unsatisfactory service. Lines typically are strung between street lampposts with ad hoc fixings and passing through trees. Reliability is poor and repair times long. Installing new wired lines is difficult because the road traffic density is so high that closing a road is problematic.

- Given that the WLL operator will provide a superior and less expensive service, it is expected to gain around 30% penetration of homes in most cities.

- In Bogota, the greater competition will reduce the available market to around 20% of all homes.

- That penetration will be reached linearly over 10 years. Such a linear increase in subscriber numbers is desirable because it allows expenditure and personnel numbers to remain relatively constant over the network lifetime.

Using the total number of homes and the expected penetration leads to the anticipated number of lines for each of the 10 key cities in Colombia as shown in Table 18.1. The table details predictions as to the total number of lines expected after 5 and after 10 years. (The numbers in Table 18.1 have been modified from the actual numbers for the system to preserve commercial confidentiality.)

18.1.2 Determining the service offering

To determine the service offering, it first is necessary to understand currently available services and expected developments in services over the life of the network. Currently, only POTS is available from the incumbent. However, it is expected that the incumbent will provide ISDN services to business in the near future. Also, in Bogota, the new WLL operators are likely to provide a wide range of services.

Taking those factors into account suggests that the minimum offering should start with ISDN capabilities and the provision of a second line to

Table 18.1
Total Market Size

Area	Anticipated Number of Lines (thousands)	
	After 5 Years	After 10 Years
Bogota	102	205
Cali	52	106
Barranquilla	20	40
Bucaramanga	8	16
Cartagena	33	66
Medellin	31	62
Santa Marta	12	24
Pereira	17	34
Villavicencio	15	29
Cucuta	11	22

subscribers. It would be wise to offer a full set of supplementary services. However, xDSL implementation is unlikely in the foreseeable future, so supporting data rates higher than the 64-Kbps ISDN B channel is probably unnecessary.

Official statistics from the incumbent operator show that at present business users generate on average 0.2E of traffic during the busy hour while residential users generate 0.09E.[1] There has been little change in those levels over the past few years, so the assumption is made that those traffic levels will remain approximately constant in coming years.

18.1.3 Traffic routing

It is important to understand the destination of calls originating in the network to be able to size interconnections between other local networks, between local and national networks, and within the network. The networks in all regions have been designed assuming the same traffic routing illustrated in Figure 18.1. Traffic routing is determined as follows:

- The incumbent publishes information showing the percentage of calls that are local and the percentage that are international. Those percentages are shown in Table 18.2 and form the basis of the calculations. It is assumed that the calling patterns will remain relatively static throughout the network lifetime.

- Based on the expected penetration, routing figures can be developed.

Table 18.2
Traffic Distribution

Local calls	93.5%
Long distance national	6%
Long distance international	0.5%

---

1. A subscriber generating 0.2E uses the phone for 20% of the time during the busy hour. Similarly, 0.09E represents 9% use, or approximately 5.5 minutes per hour.

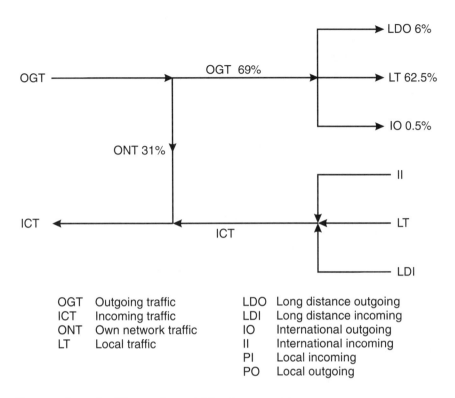

OGT	Outgoing traffic	LDO	Long distance outgoing
ICT	Incoming traffic	LDI	Long distance incoming
ONT	Own network traffic	IO	International outgoing
LT	Local traffic	II	International incoming
		PI	Local incoming
		PO	Local outgoing

Figure 18.1 Traffic routing within the network.

Taking the assumption for Bogota that 20% penetration will be achieved, which will represent one-third of the total market of 60% penetration, the routing figures are calculated as follows:

■ As described in Table 18.2, 6% of the traffic generated within the network will be passed to the national carrier and 0.5% to the international carrier.

■ Of the 93.5% of the traffic that will be local, two-thirds will be passed to other local operators; hence, only 31% of traffic generated will remain in the network.

That is represented diagrammatically in Figure 18.1, which shows that of the traffic originating within the network in Bogota, the majority (estimated at 69%) will leave the network, the remainder being destined for subscribers within the network. Of the traffic leaving the network, the majority will be local traffic destined for another operator in the same city. The remainder will comprise long distance and international traffic, which will pass to the long distance operator. The traffic routing is important in sizing the connections between the switches and in establishing interconnection agreements with other operators.

18.2 System grade of service

The *grade of service* (GOS) offered by the air interface is of prime importance in determining the cell configuration required to meet the forecast capacity demands. Setting the GOS targets requires collaboration between technical and marketing divisions. In Colombia, the WLL service will be marketed as offering a better GOS than the fixed network. Figures published by the incumbent shows that subscribers currently experience a GOS of only 99.5% (i.e., for 0.5% of the time, the telephone system is unavailable due to network faults); thus, the planned GOS for the WLL must be greater than that. Aiming for too high a GOS will result in increased network costs as additional radio channels are provided. From a marketing point of view, a GOS offering less than half the downtime of the existing networks seemed appropriate, so a GOS of 0.2% was decided on.

Table 18.3 summarizes the key objectives important in the configuration and quality of the radio part of the network. Table 18.4 addresses the GOS of the transmission and switching elements where applicable.

Table 18.3
Radio Network GOS Parameters

Parameter	Value
Air interface GOS	< 0.2%
Availability (outage planned)	99.85% (13 hours)
Availability (outage nonplanned)	99.95% (4 hours)
Dropped call (at any subscriber location)	< 1%

Table 18.4
Transmission Network GOS

Parameter	Value
Call to switch, GOS	0.1%
Switch to PSTN, GOS	0.1%

The planned outage figure is associated with the expected downtime of base stations due to software upgrades and various other network-related planned work. In general, it will be possible to organize the planned outages such that they occur at times when the network is least busy. The nonplanned outage which relates to errors in software or equipment, leads to localized unavailability for short periods of time.

18.3 Vendor selection

In Colombia, the radio spectrum available was in the 3.4 to 3.6 GHz band. Applying the filters discussed in Chapter 13 results in the short list of equipment being Proximity, Multigain, AirSpan, and AirLoop. For commercial reasons, the actual vendor selected cannot be disclosed. For the purposes of this chapter it is assumed that the Lucent system is used.

18.4 Radio Spectrum Requirements

Only the case of Bogota is considered in the calculation of the required radio spectrum. Bogota is assumed to form the worst case because it contains the highest subscriber density. The requirements for radio spectrum in the other cities generally will be slightly reduced compared to Bogota.

18.4.1 Spectrum efficiency

The AirLoop system uses a direct sequence spread spectrum carrier with a spreading bandwidth and carrier spacing of 5 MHz. The system operates within the 3.4 to 3.6 GHz fixed-services band. Two subbands are used separated by 100 MHz as shown in Figure 18.2. One band is allocated to the uplink (remote) transmitters and one to the downlink (base station); up to 10 carriers are available.

The basic cell configuration in the urban environment uses 4-by-90 degree sectors. A minimum of two carriers are required per cell in a pattern that is repeated from cell to cell, as shown in Figure 18.3. That corresponds to a frequency reuse of 2.

18.4.2 Spectrum requirements

To determine the radio spectrum requirements, it is necessary to understand the capacity of a bearer. The capacity calculations proceed as follows. First, it is important to understand that a 5-MHz bearer within the AirLoop system provides 115 channels, each of 16 Kbps. Users multiplex the required number of channels for the service they are using. For example, voice calls using the proposed 32-Kbps codec require two channels, while full ISDN service with 2B + D channels requires nine channels.

Statistics from the incumbent show that 20% of the subscribers in Bogota are business subscribers and 80% are residential. The difficulty is

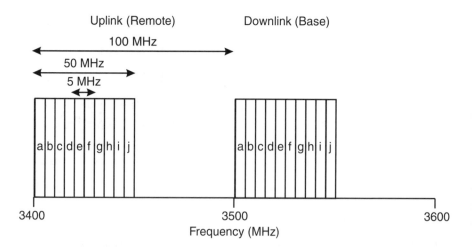

Figure 18.2 Channel plan.

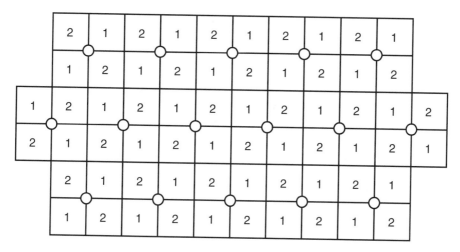

O Base Station

Figure 18.3 Sectorization.

in predicting the services that will be required. It is assumed that 80% of business subscribers will use ISDN services, and the remainder will use two-line voice-plus-data services. For the residential subscribers, it is estimated that 30% will subscribe to a second line, while the remaining 70% will take a single line. The predictions are very much approximations based on use of telephony in more developed countries and assumptions that Internet access will increase the demand for a second line among residential users. The demand for traffic channels then is given in Table 18.5.

Here, for example, the result of 0.288E for high-use business is given by $0.2 \times 0.8 \times 9 \times 0.2 = 0.288$.

This shows that the average subscriber generates the equivalent of 0.51 channel Erlangs (i.e., the user occupies a single 16-Kbps channel 51% of the time). Consulting Erlang tables, with the number of channels set at 115 and the blocking probability set at 0.2%, a total of 91E can be supported. Thus, the total number of subscribers that can be supported by each 5-MHz bearer $= 91/0.51 = 178$.

The next stage is to calculate the subscriber density and the expected cell size. Consulting a map shows that the area covered by the central part of Bogota is approximately 240 km^2. Table 18.1 showed a predicted total

Table 18.5
Demand for Traffic Channels

User	% Users	High/Low User	% Type	Channels Required	Erlangs	Total Channel Erlangs
Business	20	High	80	9	0.2	0.288
		Low	20	4	0.2	0.032
Residential	80	High	30	4	0.09	0.086
		Low	70	2	0.09	0.1
					TOTAL	0.51

of 205,000 subscribers by year 10. It is important to design the network for the final penetration levels because it is difficult to insert additional cells into the network in the same manner as for cellular systems. That is because if additional cells were inserted, subscriber units would need to be realigned, possibly mounted on different walls, and such modifications would be unacceptable to subscribers. Hence, the network design must remain stable throughout the network lifetime. Using the above figures leads to a year-10 subscriber density of approximately 850 subscribers per square kilometer.

Calculating the cell size requires an understanding of WLL propagation, a limited set of measurements, and a visual inspection of typical variations in building heights throughout the city. Bogota is composed of buildings almost universally four stories high. There are few tall buildings, except in the financial district. As a result, cellular operators typically have resorted to placing masts on top of buildings to get the antenna height some 10m above the building level. The same approach will be required for WLL, possibly using the same masts as the cellular operators. Such an approach provides a limited propagation range due to the small height increment of the base stations above the building level. Measurements from a test site showed that the typical range was approximately 1 km from the base station.

It is desirable not to reduce that range if possible, because using the maximum range achievable minimizes the amount of infrastructure required. The government had indicated that 2×50 MHz might be allocated to the operator if it could be justified. The calculation as to whether this is sufficient proceeds as follows:

- Each cell covers an area of 3.14 km^2. Each cell is divided into four sectors, with a frequency reuse of 2.

- In the area covered by one cell, $3.14 \times 850 = 2{,}670$ subscribers are expected.

- To provide service to 2,670 subscribers, $2{,}670/178 = 15$ carriers are required $= 4$ carriers per sector.

- Because of the cluster effect, each carrier requires 2×10 MHz; hence, the total spectrum requirement is 2×40 MHz.

- Allowing for some areas having a higher density of business users than others, the need for 2×50 MHz of spectrum can readily be justified.

These calculations show that, in most cases, the cell radius can be maintained at 1 km and sufficient capacity provided. It is entirely fortuitous that the system is operating at the point of just avoiding being capacity constrained.

18.5 Numbering requirements

In each city where a network is installed, it will need to be allocated an appropriate range of numbers such that calls destined for subscribers within the network can be routed to the correct switch. The size of the numbering range will need to be sufficient to cope with the expected number of subscribers, including the need for multiple numbers for some subscribers. For example, in Bogota, a numbering range providing 1,000,000 numbers (six digits) will be required. Bogota currently uses a seven-digit numbering system (an area code of 1 followed by seven digits); hence, the network would require a number of the form 6xx xxxx to accommodate subscriber growth. Alternatively, 10 sets of five digits, for example, 62x xxxx, 73x xxxx, and so on, would be required.

18.6 Network build plan

A map of Bogota is provided in Figure 18.4. Consultation with local consultants reveals that the residences unlikely to adopt a telephony service are located almost entirely on the eastern side of Bogota. Therefore, the coverage plan is to provide coverage across the entire remainder of the city. The map in Figure 18.4 shows proposed base station locations (the stars), proposed switch sites (the crosses), and links between sites (the straight lines). The sites were selected as follows:

- In the first instance, base station sites were proposed on a square grid, with base stations separated by 2 km (allowing a 1-km coverage radius).

- For each site, a survey was performed around the center of the cell. Buildings that were slightly higher than surrounding buildings, particularly if they already had masts, were spotted and, where possible, a visual inspection performed from the building.

- Because of the homogeneity of buildings throughout Bogota, in most cases it was possible to select a building very close to the proposed location.

The map also shows the interconnecting links and the locations of the switches, which are explained in Section 18.7.

18.7 Network configuration

An overview of the network is shown in Figure 18.5. Broadly, the network consists of three key elements, the base stations, the switch, and the network management system. The base stations are connected to the switch via either radio or leased lines. In turn, the switch is connected to local switches and the long-distance operator via fiber-optic cables. The network management system monitors all functions of the network and provides status reports to the network management center.

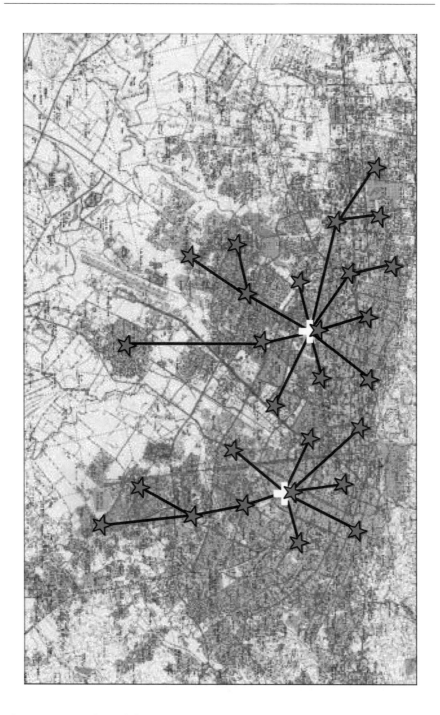

Figure 18.4 Plan of deployment in Bogota.

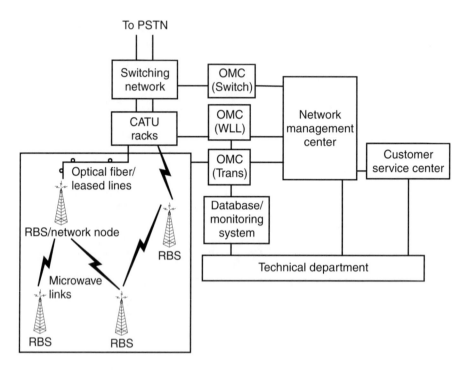

Figure 18.5 General network configuration, Bogota.

The transmission system and the switching arrangements are described in more detail next. See Section 12.2.2 for details of the AirLoop technology, including the abbreviations and nomenclature.

18.7.1 Interswitch links

The determination of the most appropriate means to connect the network switch with other local carriers and the long-distance carrier starts with a calculation of the expected traffic. In Bogota, as was shown in Table 18.1, approximately 205,000 subscribers are expected. That is actually at the upper limit of the capacity of a single switch, and it would be better to deploy two switches to handle this number of subscribers. Each switch will then handle approximately 102,000 subscribers. Assuming that each residential subscriber generates 0.09E of traffic in the busy hour, that each business subscriber generates 0.2E in the busy hour, and that the split of residential to business is 80 to 20, then the interswitch links both within the network and to external networks will require a combined capacity of:

$$(102,000 \times 0.8 \times 0.09) + (102,000 \times 0.2 \times 0.2) = 11,400E$$

Using 64-Kbps PCM, that corresponds to a total bandwidth requirement on the order of around 800 Mbps. That could be readily provided over either a fiber-optic cable or a microwave link using six 155-Mbps SDH1 links. The switches actually will be located very close to the interconnection points, so the cost of laying a cable actually is less in this case than the cost of microwave links. Further, the deployment of optical fiber allows a greater reliability and a higher capacity, should additional capacity expansion be required.

18.7.2 Interswitch signaling system

The protocol used on the interswitch links typically depends on the protocols used by the existing switches. In Bogota, because the international standard SS7 signaling system is widely used, that protocol was adopted for interconnection.

18.7.3 Interswitch synchronization

It is important that all the switches in an area are synchronized to avoid loss of data due to timing slippage. It generally is preferable to make use of existing accurate timing sources in the switches already deployed by other operators. That has the advantage of not introducing an additional timing source into the network, thus complicating the difficulties involved in interconnection. In Bogota, the incumbent already provides network timing to other local loop operators, so that timing source can be used directly.

18.7.4 Base station to switch links

In general, the connections between the base stations and the switch will be provided using microwave radios, to maximize deployment speed and minimize disruption to the city infrastructure. For each carrier deployed, an E1 link, providing 2-Mbps data capability is required on the link to the switch. On most base station sites, four sectors will be deployed, each sector supporting up to five carriers. Hence, for a fully loaded base station, around 20 E1 carriers might be required, needing 40 Mbps of data transmission capability. In practice, the microwave links likely to be deployed provide up to 16 E1 carrier capacity, that is, 32 Mbps. To avoid using two microwave links, the number of carriers deployed in those sectors of the base stations having the lowest demand will be reduced such that a 32-Mbps link can be used.

Microwave radio equipment can be supplied operating in any of the frequency bands set aside for point-to-point applications. For economic reasons, however, the 38-GHz band is preferred for links of up to approximately 6 km in length and the 23-GHz band for links up to 15 km. Almost all the links foreseen are less than 6 km in length. Because a relatively large number of base stations will be interconnected to the switch, it is not appropriate to use a ring configuration because that would result in a requirement for very high bandwidths flowing around the ring. Instead, a more appropriate configuration is to protect every microwave link, using a so-called 1 + 1 configuration. That means each link is backed up by a completely redundant link that can be activated immediately if the first link fails. Most base stations will be linked back directly to the switch in a star configuration, with some base stations daisy-chained, as was shown in Figure 18.4.

Where daisy-chaining has occurred, the connection back to the switch will require a higher bandwidth. Microwave links are widely available with 155-Mbps bandwidth (the so-called SDH1 rate), allowing the traffic from around five base stations to be concentrated onto a single link, which will be sufficient in this case.

18.8 Traffic matrix and routing

From the design principles concerning transmission described earlier and the expectations as to the amount of traffic passing to different destinations, a traffic matrix can be constructed showing how the traffic will flow around the network and through the interconnection points and showing how the required resilience is achieved within the network. The traffic matrices will be similar for all the cities, although the exact sizing of the different links may change. In Figure 18.6, the traffic matrix for Bogota is provided to illustrate the design principles used.

Figure 18.6 shows a simplified version of the Bogota network shown in detail in Figure 18.4. The two switches to be installed are shown, as are two of the switches for the other local operators and the long-distance access point. Most base stations are connected directly to the nearest switch using a 34-Mbps link. Some base stations are daisy-chained in via others. Where one base station is connected to another and then back to the switch, two 34-Mbps links are used, while where two base stations are connected to another then back to the switch, an SDH1 rate (155 Mbps) connection is the most economic way of providing the required capacity. The two switches are interconnected by a redundant 2 × SDH1 link, while the network is connected into the other switches by a ring architecture using 6 × SDH1 links. Further cross-connects can be provided in this arrangement, depending on the available infrastructure.

18.9 Summary

The network design is now complete. The switch sites have been selected and connections made to other switches and to other operators. The base

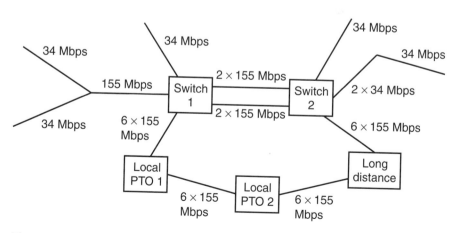

Figure 18.6 Traffic matrix for Bogota.

station sites have been selected and the means of connecting them to the switch determined. The spectrum assignment has been gained and the spectrum divided among the cell sites.

The next stage is to build the business case. That is now relatively straightforward because the following parameters have been determined:

- Anticipated number of subscribers;

- Anticipated growth in subscribers;

- Expected traffic per subscriber;

- Total equipment requirements.

Once the marketing departments have estimated the tariff levels, calculation of total revenue and total expenditure can proceed using the business case template given in Chapter 16. The full business case is not presented here due to commercial confidentiality. Suffice it to say that it was sufficiently good to proceed with the network license application.

19

The Future

THOSE WHO ARE MOST ABLE to accurately predict the future invest accordingly and make their fortunes, while those less able resort to attempting to make money from telling others what their vision of the future is. Further, most authors wisely preface predictions with some of the many examples of amusingly inaccurate predictions. Finally, anyone writing a book about a subject tends to be biased toward believing in the future potential of that subject. All those warnings are relevant here; indeed, this chapter is more at the request of the reviewers rather than the judgment of the author.

Nevertheless, predictions are interesting to read and, if treated with caution, can provide useful stimulus to thought. This chapter has been divided into four sections. The first section looks at possible technical advances in WLL technology, enabling better and lower cost offerings. The second looks at the capabilities of the other access technologies that are in competition with WLL to enhance their capabilities. The third section considers how users' demands might change in the much-heralded

"Information Society." The final section attempts to draw some conclusions from all those predictions as to how future demand for WLL systems might develop.

19.1 Technical advances in WLL

Consumers are used to computers that double in power and capacity almost annually and a seemingly invincible technical revolution. Such revolutions have not occurred in the world of wireless. Although there has been a transition from analog to digital mobile telephony, the consumer benefits have not been great, and the transition will take place over a 15-year period. Plans for third-generation mobile radio systems show little more than an ability for the phone to master multiple different standards. Equally, early predictions of the spectrum efficiency of the newer systems generally have proved to be incorrect, with GSM only providing marginal capacity increases over TACS systems.

Hence, significant technological advances enabling much more efficient use of the radio spectrum and providing valuable additional benefits to users seem unlikely. WLL systems will add additional features and gain flexibility, but in general no major breakthroughs should be expected.

It is easy to predict the arrival of broadband systems since they already are under development, which will allow WLL to enter the commercial as well as the domestic market. Along the same lines, MMDS-type systems are likely to gain useful return channel capabilities in the next few years. That would be an interesting development, since an operator deploying such a system would be able to provide both broadcast entertainment and telephony from a cost base probably only slightly greater than that for a telephony-only WLL provider. That would eventually allow a lower cost integrated service, which would seem attractive (subject to the availability of broadcast material through other media such as digital terrestrial and digital satellite broadcasting). With technical developments helping to overcome some of the effects of rain fading, these high-frequency, high-capacity systems will be highly attractive within a decade. Operators currently building lower frequency WLL networks would be well advised to bid for some MMDS spectrum to be able to partake in those developments.

By far, the most important development will be the achievement of large economies of scale through worldwide penetration, which will lead to lower prices, further increasing the cost advantage of WLL systems. At present, subscriber units cost perhaps $1,000 for the proprietary systems, whereas GSM mobile phones cost around $200 each. In principle, there is little reason why WLL subscriber unit prices should not fall to the level of mobile phones, or near those levels, as long as sufficient economies of scale can be achieved. That is a reinforcing trend; as prices fall, demand increases, resulting in greater economies of scale and further reductions until the equipment cost falls to the minimum manufacturing cost. WLL will undoubtedly follow that path since it already has taken the first steps along it, and orders are increasing even at current prices. However, it will not follow it so quickly as the mobile systems due to the lack of standardization and resulting lack of competition. Nevertheless, within a decade, subscriber units should cost less than around $400 each in real terms.

19.2 Technical advances in other access techniques

The key question facing WLL systems is whether the xDSL technologies will develop to such a level that subscribers will be able to cheaply receive VOD and similar material through their existing copper pairs. Such a development would allow the wired operators to subsidize their cost of telecommunications with additional services, making market entry difficult for WLL operators. The success of xDSL depends almost entirely on the quality of the existing lines, which varies from region to region and country to country. Prediction in this area is almost impossible; however, it seems likely that the cost of xDSL provision will not fall to the level where it becomes attractive for mass-subscription use, such as VOD. Instead, it will be of most interest to a limited number of people working from home and willing to pay perhaps as much as $1,000 in total for connection. That limited market will not achieve the economies of scale necessary to drive the prices down, so xDSL will remain a niche application, not seriously threatening WLL operators.

Another important issue is whether the much-heralded fixed-mobile integration will occur. Perhaps integration is a misnomer; mobile dominance might be a better way to describe the vision of those articulating such a future. Simply put, why not use your mobile at home as well as when roaming? As long as the quality is sufficient and the call charges are low, that makes establishing contact with an individual much simpler. As discussed at length in Chapter 15, there are many flaws with that argument, flaws that are increasingly emerging with time. Fixed-mobile integration will not occur in any meaningful way over the next decade. Instead, intelligent network platforms will perform call redirection between fixed and mobile phones (where the fixed phone might actually be a home cordless base station), depending on a wide range of options so that single numbers can be used and single bills provided. Indeed, such platforms are already here and are used for premium "personal numbers" in some countries. Such developments hold no concerns for WLL; a fixed line to the home will still be required.

Internet telephony currently is an application that concerns many existing operators. The concept is to use advanced speech coders to compress and segment the voice and then send the packets over the Internet. The advent of the technology is entirely artificial due to the fact that all Internet calls are charged at local rates despite the fact that many result in international flows of data. That, in fact, reflects the true cost of telephony provision, where the local loop section represents the vast majority of the cost and the national and international trunk contribution is almost insignificant. However, fixed operators persist in inflating international prices to increase profitability and in some cases subsidize the local loop. If Internet telephony works (and it is unlikely to for long, even if it does work at first, as the network becomes swamped by voice as well as the current heavy data load), the net result simply will be that fixed operators will rebalance their tariffs so that Internet telephony is no longer worthwhile. The only advantage then will be the ability to link voice with exploring a site so that users could contact the operator of the site they are surfing for help.

Even that discussion is irrelevant for WLL. Whether the voice service is through the voice network or across the Internet, it still will be carried along the same WLL lines to the network and can still be tariffed at the same rate. For that reason, Internet telephony holds little threat for the WLL operator.

19.3 Changing user demand

It often is said that the world is becoming the Information Society, with huge amounts of information being made available to individuals and passed around the world as rapidly as required. The Internet is held up as the first manifestation of such a development. If true, users will increasingly demand bandwidths of megabits per second as services are provided that require them. Certainly, the Internet is increasingly requiring high bandwidth access, and as file sizes and hard disks get increasingly large, it is undeniable that user demand for data is increasing.

It is easy to get carried away with this approach. Most people rely on broadcast entertainment, which will provide more and more channels and accompanying data. That data will be delivered via the broadcast, not the telephony channels. Those who require large files will require them only periodically. Finally, and most important, users will want such services only if they can be provided at a low price; for example, most Internet users could benefit from an ISDN connection but do not choose to have one because the additional cost exceeds the value that an ISDN line would add for them.

Therein lies the problem. Ubiquitous provision of high-bandwidth services will require enormous upgrade to the world's telecommunications network, particularly switches and packet servers and certain parts of the interconnecting network, not to mention the access network. Such upgrades cost money, and as yet few users value the Information Society enough to pay the additional cost required. Within the next 10 to 20 years, there may be many users who would like a high-speed connection but few who are prepared to pay for it. The user demand will be for information at a reasonable cost that will encourage information providers to compress data and to use graphics more sparingly.

Indeed, perhaps bizarrely, at the moment the user is actually prepared to pay less for higher bandwidth services. Figure 19.1 shows approximate current levels paid per minute for a range of different services by users in developed countries.

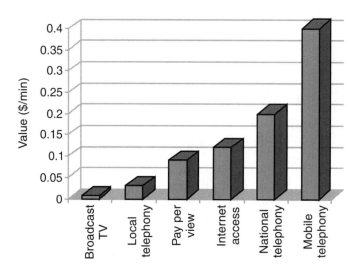

Figure 19.1 Variation in value for different services.

As can be seen in Figure 19.1, the highest value services typically are also the most narrowband. If the bandwidth per service is factored into the analysis to present value per minute per megahertz, the graph becomes even more extreme; indeed, the variation is so great it must be shown in a table, as shown in Table 19.1.

What Table 19.1 shows is that not only are users not prepared to pay more for broadband services, they are not even prepared to pay as much; typically they expect to pay something like one-thousandth the amount on a comparable per-megahertz basis for broadband services. Because the

Table 19.1
Relative Value Placed on Narrowband and Broadband Services

Service	Service Type	Bandwidth	Value/min/MHz
Broadcast TV	Broadband	8	0.000875
Pay-per-view	Broadband	8	0.01125
Local telephony	Narrowband	0.025	1.2
Internet access	Narrowband	0.025	4.8
National telephony	Narrowband	0.025	8
Mobile telephony	Narrowband	0.025	16

cost of provision is approximately linear with the amount of bandwidth (since the bandwidth could be used for multiple narrowband services rather than one wideband service), there is a massive gulf between cost of provision and expectation of cost. That will prevent the widespread adoption of broadband services for many years to come.

Although the demand for high-bandwidth provision will not rise dramatically, the use of the access network for applications other than voice will change significantly. Internet access, including applications such as home shopping, will increase dramatically the traffic generated by a home in that the phone line will now be used a much greater percentage of the day. Modems will become a standard fit in computers and probably also in televisions. That will fuel the demand for two or more lines per house and will also increase the revenue received per line by the operator.

A final area worthy of mention is the increased environmental awareness, which will make the laying of additional cables increasingly problematic. Even now, cable operators are criticized for damaging tree routes. Once they are forced to pay compensation for the inconvenience their digging causes, wireless will become the preferred option for most connections.

19.4 The future for WLL

Of course, this book would not have been written if it was not thought that WLL will have a bright future. That view is shared by most industry experts, as is shown in the predictions in Figure 19.2 (reproduced from Chapter 5).

Recall also the prediction made by analysts that by the year 2000 over 10% of all lines being installed will be wireless. Other analysts have predicted that the number of wireless lines installed per year could overtake wired lines before the year 2005.

Such a prediction is not difficult to believe when the following factors are taken into consideration:

- The cost per line for WLL currently is lower than for wired systems and will fall further with economies of scale.

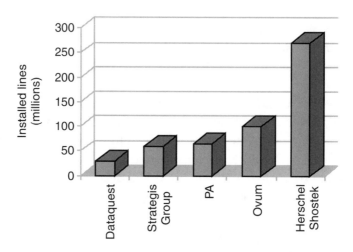

Figure 19.2   Market predictions for WLL made at the start of 1997.

- The demand for telecommunications around the world will increase significantly in coming years.

- Fragmentation and the development of market niches in the telecommunications market will allow a range of different operators to coexist.

The predictions do not hinge on any technical or commercial breakthrough, although they do require that other access technologies do not develop unexpectedly or that other access technologies are discovered.

Of course, there will be some problems. Some operators will fail in the first few years through a misunderstanding of the market or through the deployment of inappropriate equipment. Problems soon will emerge with lack of radio spectrum, and that, in effect, will increase the cost of radio spectrum, either directly through auctions or indirectly through the impossibility of obtaining more, forcing greater network costs as smaller cells are deployed. That will be partially solved as higher frequencies are used, but as with mobile radio, shortages of spectrum will remain a permanent problem, limiting subscriber numbers and the data rates that can be offered.

The future of WLL is bright. Hopefully, you are now persuaded to invest some of your time or money in it.

List of Acronyms

ADPCM	adaptive differential pulse code modulation
ADSL	asymmetric digital subscriber line
AM	amplitude modulation
AMPS	Advanced Mobile Phone System
ARQ	automatic repeat request
BCH	Bose-Chaudhuri-Hocquenghem
BER	bit error rate
BRA	basic rate access (ISDN at 144 Kbps)
BSC	base station controller
CAS	channel associated signaling
CATU	central access and transcoding unit
CDMA	code division multiple access

CEPT	European Conference of Postal and Telecommunications Administrations
CT-2	Cordless Telephony system 2
CTRU	central transceiver unit
D-AMPS	digital AMPS
DAN	DECT access node
DCA	dynamic channel allocation
DCS1800	digital cellular system at 1800 MHz
DECT	digital enhanced cordless telephone
DMT	discrete multitone
DPSK	differential phase shift keying
DS	direct sequence
DSL	digital subscriber line
ERMES	European Radio Messaging System
ETSI	European Telecommunications Standards Institute
FAU	fixed access unit
FDM	frequency division multiplexing
FDMA	frequency division multiple access
FH	frequency hopping
FTTC	fiber to the curb
FTTH	fiber to the home
FRA	fixed radio access
GAP	generic access protocol
GDP	gross domestic product
GMSK	Gaussian minimum shift keying
GSM	global system for mobile communications
HDSL	high-speed digital subscriber line
HDTV	high-definition TV
HFC	hybrid fiber coax
IRR	internal rate of return
ISDN	Integrated Services Digital Network

ISI	intersymbol interference
ISM	industrial, scientific, and medical
ITS	intelligent telephone socket
ITU	International Telecommunications Union
LAN	local area network
LMDS	local multipoint distribution systems
LOS	line-of-sight
MAN	metropolitan area network
MMDS	microwave multipoint distribution system
MVDS	microwave video distribution system
NEXT	near-end cross-talk
NIU	network interface unit
NMT	Nordic Mobile Telephone
NPV	net present value
OFDM	orthogonal frequency division multiplexing
O&M	operations and maintenance
PABX	private access branch exchange
PAMR	public access mobile radio
PCM	pulse code modulation
PCS	personal communications service
PHS	personal handiphone system
PM	phase modulation
PMR	private mobile radio
PN	pseudo-random noise
POTS	plain old telephony service
PSTN	public switched telephone network
PTO	Post and Telecommunication Organization
QAM	quadrature amplitude modulation
QPSK	quadrature phase shift keying
RBS	radio base station

RF	radio frequency
RFA	radio fixed access
RLL	radio in the local loop
RNC	radio node controller
RPE-LTP	regular pulse excited-long term prediction
RPI	retail price index
RS	Reed Solomon
SIR	signal-to-interference ratio
SMR	specialized mobile radio
SNR	signal-to-noise ratio
STRU	subscriber transceiver unit
TACS	total access communications system
TCM	trellis code modulation
TDD	time division duplex
TDMA	time division multiple access
TETRA	Trans-European Trunked Radio
UHF	ultra high frequency
USO	universal service obligation
VDSL	very high-speed digital subscriber line
VSAT	very small aperture terminal
VOD	video on demand
WiLL	Motorola's version of wireless local loop
WLL	wireless local loop
xDSL	belonging to the family ADSL, HDSL, and VDSL

About the Author

WILLIAM WEBB graduated in electronic engineering with a first class honors degree and all top prizes in 1989. In 1992 he received his Ph.D. in mobile radio, and in 1997 he was awarded an M.B.A., all from Southampton University, United Kingdom.

From 1989 to 1993, Dr. Webb worked for Multiple Access Communications Ltd. as Technical Director in the field of hardware design, modulation techniques, computer simulation, and propagation modeling. In 1993 he moved to Smith System Engineering Ltd., where he was involved in a wide range of tasks associated with mobile radio and spectrum management. In 1997 he joined Netcom Consultants, where he is a principal consultant in the Wireless Access division.

Dr. Webb has published more than 40 papers, holds four patents, was awarded the Institute of Electrical and Radio Engineers Premium in 1994, and is the co-author, with L. Hanzo, of *Modern Quadrature Amplitude Modulation* (New York: Wiley & Sons, 1992). He is a member of the IEE and a senior member of the IEEE.

Index

The Artech House Mobile Communications Series

John Walker, Series Editor

Mobile Information Systems, John Walker, editor

Personal Communications Networks, Alan David Hadden

RF and Microwave Circuit Design for Wireless Communications,
Lawrence E. Larson, editor

Smart Highways, Smart Cars, Richard Whelan

Spread Spectrum CDMA Systems for Wireless Communications,
Savo G. Glisic, Branka Vucetic

Transport in Europe, Christian Gerondeau

Understanding GPS: Principles and Applications, Elliott D. Kaplan, editor

Vehicle Location and Navigation Systems, Yilin Zhao

Wireless Communications for Intelligent Transportation Systems,
Scott D. Elliott, Daniel J. Dailey

*Wireless Communications in Developing Countries: Cellular and
Satellite Systems,* Rachael E. Schwartz

Wireless Data Networking, Nathan J. Muller

Wireless: The Revolution in Personal Telecommunications, Ira Brodsky

For further information on these and other Artech House titles,
including previously considered out-of-print books now available
through our In-Print-Forever™ (IPF™) program, contact:

Artech House	Artech House
685 Canton Street	Portland House, Stag Place
Norwood, MA 02062	London SW1E 5XA England
781-769-9750	+44 (0) 171-973-8077
Fax: 781-769-6334	Fax: +44 (0) 171-630-0166
Telex: 951-659	Telex: 951-659
e-mail: artech@artech-house.com	e-mail: artech-uk@artech-house.com

Find us on the World Wide Web at: www.artech-house.com